XR Haptics: Implementation & Design Guidelines

with Enterprise VR application areas, use cases, and implementation examples

Authors:

Eric Vezzoli, Chris Ullrich, Gijs den Butter, Rafal Pijewski

Contributors:

Pf. Vincent Hayward,
Sorbonne University

Pf. Ed Colgate,
Northwestern University

Pf. Allison Nakamura,
Stanford University

Dr. Jess O'Brian,
Meta Reality Labs

Dr. Oliver Schneider,
University of Waterloo

Dave Birnbaum,
Immersion Corporation

Dr. William Frier,
Ultraleap

Joe Michaels,
Haptx

Daniel Shor,
Contaxtual Labs

Dr. Margot Racat,
IDRAC Business School

Copyright © 2022 by Haptics Industry Forum

www.hapticsif.org

All Rights Reserved.

ISBN: 978-0-578-39120-5

No part of this work may be reproduced, distributed, or transmitted in any form or by any means, including photocopying, recording, or other electronic or mechanical methods, or by any information storage and retrieval system without the prior written permission of the author(s), except in the case of brief quotations embodied in critical reviews and certain other noncommercial uses permitted by copyright law.

Introduction

To emphasize the importance that touch has in our perceptual system, think of an action we perform every day, such as pouring water into a glass. Now let's imagine performing this action without haptic perception. We would not be able to tell when we firmly grasp the bottle, and we would not be able to precisely control the movement of the bottle as we pour the water. In other words, without the sense of touch, it would be very difficult, if not impossible, to perform manipulation tasks like the one I just described. This example highlights why experiences developed in virtual contexts, such as metaverses, are still far from being truly engaging. Tactile feedback technologies are absolutely necessary.

Tactile feedback in extended reality is very important not only to perform an action such as pouring water but also, for example, to perceive the pleasure of fabrics under our fingertips or to improve the sense of body ownership and presence of in the metaverses. This explains why users experimenting with virtual and extended reality applications very often complain of a lack of tactile sensation.

Today there are mass-market solutions that offer rather simple tactile feedback, such as Oculus Rift 2, that manage to give only a vague idea of the interaction with virtual environments and are very far from creating truly immersive experiences. Indeed, several interfaces have been developed so far, but there are still no XR applications that provide a level of tactile realism and engagement comparable to, for example, that of auditory or visual feedback generated by headphones or visual displays. This is proof that further research and more integration efforts are needed.

Furthermore, the complexity of the technologies adopted to provide the user with a certain level of tactile realism is generally difficult to reconcile with multiple application design guidelines.

The current situation is clear and paradoxical. We have technologies for developing immersive XR applications and technologies for creating realistic tactile sensations, but we still lack a clear set of guidelines and best practices to develop and integrate them.

XR Haptics is a great book that addresses the need to systematically introduce technologies and methodologies of haptic feedback into mixed reality projects. The book is very well written and organized and provides a comprehensive introduction to the use of haptics in extended reality.

To provide a guide to orient implementers, designers, and developers working in extended reality applications, the book features best practices and tips learned from veteran haptic experts. Following a clear and smooth path, the author guides readers towards the idea that haptics represents tangible value for today's real-world use cases in extended reality. Developers and professionals can rely on this book to have a how-to manual for tackling most problems related to introducing haptics in extended reality.

The book consists of two main parts. In the former, the authors carefully drive the reader into the haptic world, describing all the aspects that have to be considered in developing and integrating haptics into XR projects. Several concepts were discussed, ranging from haptic technologies and pseudo-haptics to development frameworks.

In the last part, the focus is on applications. Readers can find useful information for their own projects explained through a list of use cases with a critical discussion.

XR Haptics is a book that all beginners, professionals, and more experienced users should read for support in developing and improving their XR projects with haptics.

Domenico Prattichizzo

University of Siena and Istituto Italiano di Tecnologia

EIC IEEE Transactions on Haptics

Table of Contents

1. Overview of the book and How to Use it 11
2. What is XR haptics? 13
3. How to implement XR haptics 14

Part I: Haptic Integration Framework 16

4. Multisensory Requirements 16
   4.1. Overview 16
      4.1.1. Is there a role for haptics in the simulation? 16
      4.1.2. How will haptics relate to other available modalities? 17
5. The Goal of Haptics in the Scenario 19
   5.1. Overview 19
   5.2. Actions Required 22
   5.3. Key decision to make 23
   5.4. Key Considerations 23
   5.5. Deliverables 23
6. Haptic Interaction Design 24
   6.1. Overview 24
   6.2. Haptic Interactions 24
   6.3. Haptic Interaction-Objective Considerations 30
   6.4. Actions Required 32
   6.5. Deliverables 33
7. Hardware device choice 34
   7.1. Overview 34
   7.2. Haptic Feedback Taxonomy 34
      7.2.1. Non-Spatial - Vibrotactile 36
      7.2.2. Kinesthetic - Resistive Force Feedback 37
      7.2.3. Kinesthetic - Active Force Feedback 37
      7.2.4. Contact Spatial - Surface Friction 38
      7.2.5. Contact Spatial - Skin Indentation 38
      7.2.6. Contact Spatial - Electrostimulation 38
      7.2.7. Non-Contact Spatial - Ultrasound 39
   7.3. Technology – Interaction considerations 39

- 7.4. Related Haptic Solutions ... 41
  - 7.4.1. Passive Haptics Considerations ... 41
  - 7.4.2. Pseudo Haptics ... 43
  - 7.4.3. Self-Referenced Haptics ... 44
- 7.5. Actions Required ... 45
- 7.6. Deliverables ... 46
- 7.7. Key Considerations ... 46

8. Haptic Development Environment ... 48
- 8.1. XR Development Frameworks ... 48
  - 8.1.1. Unity ... 48
  - 8.1.2. Unreal ... 49
  - 8.1.3. OpenXR ... 50
  - 8.1.4. Device-Specific SDKs and Frameworks ... 50
  - 8.1.5. Generic Haptic SDKs and Frameworks ... 51
- 8.2. Haptic Effect Taxonomy ... 52
  - 8.2.1. Static vs. Dynamic ... 52
  - 8.2.2. Vibration Effects – static and dynamic effects ... 53
  - 8.2.3. Kinesthetic Effects – dynamic effects ... 57
  - 8.2.4. Non-Contact Spatial – static and dynamic effects ... 59
- 8.3. Implementation considerations ... 59
- 8.4. Encoding Standardization of Haptics Data Considerations ... 60
- 8.5. Actions Required ... 61
- 8.6. Deliverables ... 61

9. 3D Design Import, Haptics, and Interaction Design ... 62
- 9.1. Overview ... 62
- 9.2. 3D Design Import ... 62
- 9.3. Haptics-Interaction Framework ... 63
  - 9.3.1. Egocentric Haptics ... 64
  - 9.3.2. Allocentric Haptics ... 65
- 9.4. Interaction Loop ... 66
  - 9.4.1. Collision detection ... 66
  - 9.4.2. Internal Dynamics ... 67
  - 9.4.3. Events (non-diegetic) ... 68
  - 9.4.4. Haptics rendering and mixing ... 68
- 9.5. Actions Required ... 69

9.6.	Deliverables	70
10.	Multimodal Integration	71
10.1.	Overview	71
10.2.	Implementation Guidelines	73
10.2.1.	Congruency Guidelines	73
10.2.2.	Synchronicity guidelines	74
10.2.3.	Multimodal Implementation example	74
10.3.	Actions Required	75
10.4.	Deliverables	75
11.	Haptic design framework for XR	76
11.1.	Example scenario	76
11.2.	Chapter 4 deliverable: creating your XR scenario	80
11.3.	Chapter 5 deliverable: Defining the haptic goals	80
11.4.	Chapter 6 deliverable: Sketch haptic interactions	81
11.5.	Chapter 7 deliverable: Haptic technology map	82
11.6.	Chapter 8 deliverable: Haptic design choice	83
11.7.	Chapter 9 deliverable: Prototype interactions	84
11.8.	Chapter 10 deliverable: Integrated multi-sensory scenario	85
11.9.	Testing and Iteration	86
12.	Upgrading and Maintenance	86
Part II: Use Cases and Applications		88
13.	Enterprise XR Application Areas	89
13.1.	Application Area: Virtual Prototyping	89
13.1.1.	Overview	89
13.1.2.	Investment Range	89
13.1.3.	Non-Haptic Considerations	89
13.1.4.	Customer Goals	90
13.1.5.	Role of Haptics	90
13.1.6.	Haptic Considerations for Success	91
13.1.7.	Use Cases	91
13.2.	Application Area: Training	92
13.2.1.	Overview	92
13.2.2.	Investment Range	92
13.2.3.	Non-Haptic Considerations	92

- 13.2.4. Customer Goal(s) 92
- 13.2.5. Role of Haptics 93
- 13.2.6. Haptic Considerations for Success 93
- 13.2.7. Use Cases 94

13.3. Application Area: Marketing/Sales 95
- 13.3.1. Overview 95
- 13.3.2. Investment Range 95
- 13.3.3. Non-Haptic Considerations 95
- 13.3.4. Customer Goals 95
- 13.3.5. Role of Haptics 96
- 13.3.6. Haptic Considerations for Success 96
- 13.3.7. Use Cases 97

13.4. Application Area: Tele-existence/Tele-robotics 98
- 13.4.1. Overview 98
- 13.4.2. Investment Range 98
- 13.4.3. Customers Goals 98
- 13.4.4. Role of Haptics 99
- 13.4.5. Haptic Considerations for Success 99
- 13.4.6. Use Cases 99

13.5. Application Area: Assistive 100
- 13.5.1. Overview 100
- 13.5.2. Investment Range 100
- 13.5.3. Customer Goals 100
- 13.5.4. Role of Haptics 100
- 13.5.5. Haptic Considerations for Success 101
- 13.5.6. Non-Haptic Considerations 102
- 13.5.7. Use cases 103

14. Use Cases 104

14.1. Electrical Maintenance Training 104
- 14.1.1. Overview 104
- 14.1.2. Customer Goals 105
- 14.1.3. Project Budget 105
- 14.1.4. Haptic Technologies Used 105
- 14.1.5. Role of Haptics 105
- 14.1.6. Outcome 105

14.2. Automotive Painting Training 105
- 14.2.1. Overview 105
- 14.2.2. Customer Goals 106
- 14.2.3. Project Budget 106
- 14.2.4. Haptic Technologies Used 106
- 14.2.5. Role of Haptics 106

14.2.6.	Outcome	106
14.3.	**Satellite Assembly Training**	**107**
14.3.1.	Overview	107
14.3.2.	Customer Goals	107
14.3.3.	Project Budget	107
14.3.4.	Haptic Technologies Used	107
14.3.5.	Role of Haptics	107
14.3.6.	Outcome	108
14.4.	**Discover History with Haptics**	**108**
14.4.1.	Overview	108
14.4.2.	Customer Goals	108
14.4.3.	Project Budget	108
14.4.4.	Haptic Technologies Used	109
14.4.5.	Role of Haptics	109
14.4.6.	Outcome	109
14.5.	**Electrical Assembly Training**	**109**
14.5.1.	Overview	109
14.5.2.	Customer Goals	110
14.5.3.	Project Budget	110
14.5.4.	Haptic Technologies Used	110
14.5.5.	Role of Haptics	110
14.5.6.	Outcome	110
14.6.	**Nerve damage experience**	**110**
14.6.1.	Overview	110
14.6.2.	Customer Goals	111
14.6.3.	Project Budget	111
14.6.4.	Haptic Technologies Used	111
14.6.5.	Role of Haptics	111
14.6.6.	Outcome	111
14.7.	**Augmented Reality Haptics Showcase**	**112**
14.7.1.	Overview	112
14.7.2.	Customer Goals	113
14.7.3.	Project Budget	113
14.7.4.	Haptic Technologies Used	113
14.7.5.	Role of Haptics	114
14.7.6.	Outcome	114
14.8.	**Medical Care Training**	**114**
14.8.1.	Overview	114
14.8.2.	Customer Goals	115
14.8.3.	Project Budget	115
14.8.4.	Haptic Technologies Used	115

	14.8.5.	Role of Haptics	115
	14.8.6.	Outcome	115
15.	**Technology Options and Implementation Examples**		**116**
	15.1.	Overview	116
	15.2.	Military Training	116
	15.3.	Procedural Training	118
	15.4.	Telerobotic	120
	15.5.	Virtual Prototyping	122
	15.6.	Marketing	124
16.	**Conclusions**		**126**
17.	**Acknowledgements**		**126**
18.	**Authors Biography**		**127**
19.	**Glossary**		**130**
20.	**Works Cited**		**134**

1. Overview of the book and How to Use it

This book is intended to orient implementers, designers, product, and program managers working in extended reality application areas who are considering adding haptic feedback technologies to their new or existing projects. It collects several decades of combined experience in this specific area by members of the Haptics Industry Forum [1], along with tips and suggestions from world leaders in haptics.

Incorporating haptic feedback into an extended reality (XR) project can dramatically impact the user's sense of presence and agency, which increases training value and engagement with the XR experience. The key challenge that many developers encounter when incorporating haptics is the immaturity of tooling, lack of a universal SDK, and critical design, implementation, and playback differences from audio and visual experience components. This may seem daunting, but with the correct perspective and expectations, haptics can generate real value-add without derailing project timelines or budgets. This book aims to provide a practical guide for achieving the promise of haptics in XR projects.

First and foremost, this book is a pragmatic guide designed to enable real-world outcomes. It includes a step-by-step guide for implementers to approach the implementation of haptics in their scenario in a controlled way with clear milestones, deliverables, and risks along the way. Even if the process suggested by this book is not strictly followed, the basic guidelines provided about haptic development may be utilized throughout the implementation process to achieve a more successful outcome.

The approach outlined in this book aims to provide a global taxonomy of the so-called "Haptics-Interaction problem" in extended reality. Implementers tasked with developing and integrating haptics into an XR project will find the section-by-section guide for the haptics-specific implementation, along with decisions, deliverables, and risks for each step. Implementers should start with the workflow proposed in Part 1 and dive into the details of each step in Chapters 4 through 12.

In Part 2, Product and Program managers will find XR use cases in which haptics are known to provide meaningful value. A selection of real-world case studies is presented in Chapters 13 and 14.

Tips from world-class haptics experts are inserted throughout the book. These serve to guide the reader with concise rules of thumb that were learned from years of experience developing haptic technology. The effort has been made to illustrate essential concepts with diagrams and references to external resources.

The reader is encouraged to approach this book first and foremost from a design perspective. A key challenge for haptic experiences is that certain decisions are costly and expensive to recover from. For this reason, careful design and planning are needed before a single decision is made. The effort should be primarily driven by carefully constructed goals which imply interactions, haptic effects and ultimately software and hardware choices. This is illustrated in Figure 1 below and readers are strongly advised not to skip steps and not to start implementation prematurely.

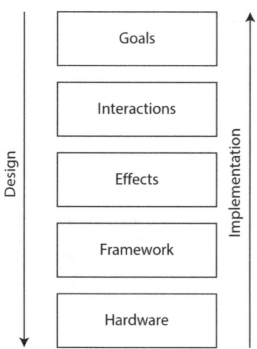

Figure 1: Suggested design driven approach to use this book.

The recommended way to read this book is to first read through to at least Part 1, but ideally the entire book. This will provide the project team with a good overview of all the considerations, issues, and opportunities for haptics. Once that is complete, we recommend using the worksheet example in Part 2 and filling this out as the team works through each chapter during the integration process.

2. What is XR haptics?

First and foremost, the term haptics refers to technologies intended to provide touch feedback to users. Haptic devices are increasingly common among consumer electronics, game controllers, and professional devices for VR, AR, telerobotics, and many other application areas.

XR haptics refers to the inclusion of some type of touch stimulation system in the context of an eXtended Reality (XR) experience. Although there are mass-market solutions that offer primitive touch feedback (e.g., Oculus Rift 2), most XR use cases for haptics today are within the enterprise domain. The primary reasons for this are the cost and complexity associated with the electromechanical stimulation systems necessary for haptic feedback. That said, high-quality haptic devices are rapidly becoming affordable across many industries, prosumer applications, and the premium mass market. End users at all levels have been exposed to increasingly high-quality haptics and realize its value. This guide will extract many of the insights gained in implementing enterprise VR haptics to enable haptics to be integrated at any price point.

Throughout this text, the term XR is intended to be interchangeable with various flavors of immersive experiences, including virtual reality (VR), augmented reality (AR), and mixed reality (MR).

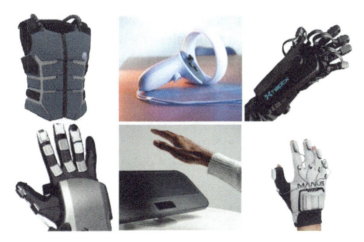

Figure 2: Example haptic devices typically used in XR haptics. Clockwise from upper-left is the Skinetic vest from Actronika, Oculus Quest 2 controller from Meta, HaptX Developer Kit from HaptX, the Manus prime II Haptics from Manus, Stratos from Ultraleap, and Nova from Senseglove.

3. How to implement XR haptics

This book is divided into two Parts which broadly cover the implementation of haptics and haptic application areas, respectively. The first part is intended to be a step-by-step guide for practitioners. The section part is intended to support the planning of new XR projects that may incorporate haptics.

Part I: Haptic Integration Framework

Successfully incorporating haptics into a new or an existing project is best achieved by following a systematic process like the one outlined in Figure 3. This process is used as a framework for the remainder of the text and is driven by a series of questions:

- Chapter 4: What modalities will be present, and how should they relate?
- Chapter 5: What are the specific goals to be achieved by haptics within the product?
- Chapter 6: What are the interaction mechanics for each haptic goal?
- Chapter 7: What haptic hardware can fulfill the haptic goals and interaction mechanics?
- Chapter 8: What development environment and runtime engine will be used to develop the haptic feedback?
- Chapter 9: How will the haptic sensations be designed?
- Chapter 10: How will haptics work with audio and visual simulation components?
- Chapter 11: A worked example of the activities of chapters 4-10 for a representative use case.
- Chapter 12: What special considerations exist for an upgrade scenario?

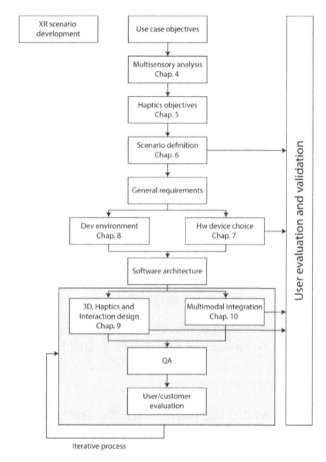

Figure 3: Implementation process for a new XR system with haptics. The numbers on the blocks indicate the related chapter.

Part II: Use Cases and Applications

Chapters 13-15: These chapters provide use case and application area detail to help inform practitioners of the various use cases and considerations for integrating haptics in a wide variety of real-world scenarios.

As with any user-facing product or experience, it is critical to incorporate customer, stakeholder, and user feedback into the process as early and often as possible. In particular, it is recommended that an iterative, user-centric approach be utilized to ensure that the haptic design maximizes its value relative to the overall product.

Part I: Haptic Integration Framework

The first portion of this book is a step-by-step guide intended to be a convenient and hands-on manual for evaluating, designing, and implementing haptics in an XR system.

4. Multisensory Requirements

4.1. Overview

In nearly every XR simulation, haptic feedback will need to work in concert with visual and audio feedback. The different qualities and values of each sensory modality are the starting point for determining the haptic functionality of the simulation. For this reason, this initial chapter presents a structured method for determining what the role of haptics will be and how it will relate to other sensory stimuli.

It is recommended that practitioners first review this entire guide and then return to this section to address the specific goals of haptics. Next, it is recommended that the development team and stakeholders honestly answer the following questions.

4.1.1. Is there a role for haptics in the simulation?

Haptic feedback is not a panacea for realism or presence in a simulation. Stakeholders should carefully consider what role haptic feedback is intended to have. The common objectives of haptics are discussed in Chapter 5 and should be thoughtfully discussed and prioritized. It is important to acknowledge that haptics does not add value to every simulation. In some cases, haptic feedback can be more distracting or confusing than no feedback at all.

Readers are encouraged to carefully consider and discuss the goals of adding haptic feedback to a simulation. Some possible prompts include:

1. How important is touch to the real-world analog of the simulation?
2. Will haptics create training value?
3. Will haptic feedback meaningfully increase user engagement?
4. How will users value and perceive the presence/absence of haptic feedback?

Stakeholders should discuss these issues before significantly investing in haptic design or functionality.

4.1.2. How will haptics relate to other available modalities?

Humans create their perception by fusing sensory information from all available senses. Adding haptics to a simulation without considering its relationship to audio and visual stimuli is a recipe for disaster. The best practices for multimodal integration are discussed in Chapter 10 and should be reviewed and agreed upon by stakeholders and the product team.

Some possible discussion topics include:

1. Can training feedback be provided by audio and/or visual cues alone?
2. What is the desired relative fidelity of visual, audio, and tactile feedback? Are these levels sufficiently similar?
3. Does the simulation system or game engine permit synchronized multi-modal rendering?

The answers to these questions can serve as guiding principles by which key technical decisions may be made. Readers are also suggested to review the application areas outlined in Chapter 13, which contain multiple examples of successful multimodal integration use cases.

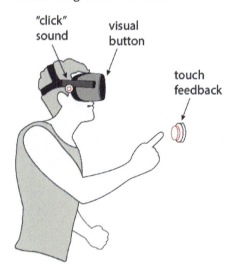

Figure 4: A multisensory haptic virtual interaction

Tip from the expert:

Sketch out your interactions over the story of the experience. Define the goal for each interaction based on the end-user's input before you dive into programming.

Dr. Oliver Schneider, assistant professor University of Waterloo.

This chapter does not have precise deliverables. Instead, it describes how a holistic reflection upon a project plan can help determine if and where haptics can bring value.

5. The Goal of Haptics in the Scenario

5.1. Overview

The first step is to clearly and carefully define the goals or objectives that haptics will enable in the scenario. It is crucial to identify the purpose and role of haptics in the XR scenario since haptic feedback provides an additional channel of information to the user. Typically, this step will involve collaboration with the end-user and the product manager and may be iterative based on findings from prototyping activities and user feedback.

Taking time to evaluate and articulate the haptic feedback goals formally is a critical step. Unlike audio and video, haptic devices, in general, cannot render all possible touch sensations, and trade-offs must be made during hardware selection.

The goal of this chapter is to guide the product team to explicitly identify and prioritize the simulation goals associated with haptics so that hardware selection can support these goals rather than restrict what's possible. (Hardware choices are discussed in Chapter 7.)

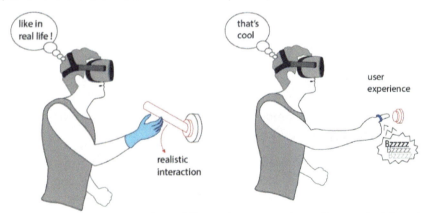

Figure 5: Representation of two different haptics goals, realism on the left and user experience on the right.

Below is a list of haptics' goals in a virtual experience and are not listed in any particular order. For implementors, this list can serve to facilitate communication and requirements gathering with stakeholders. Note that the listed objectives can be referenced to the Role of Haptics in each Application Area presented in Chapter 0.

Realism: Realism is the idea that perception in a virtual environment is similar to that of a real environment. This is also known as "content validity." [23] For haptics, the feedback provided during contact or interaction is as similar as possible to the comparable physical touch experience. Except for a very narrow set of sensations, it is expensive and complex to achieve a semblance of haptic realism. For this reason, it's important to carefully understand and align with stakeholders on which specific elements in the virtual environment need to have haptic realism, if any.

Immersion: Immersion refers to enhancing a user's sense of presence. Immersion results when a scenario behaves in a *plausible* way, and when virtual events and objects have associated visuals, sounds, and haptics that are consistent with each other.

Note the subtle but important difference between Immersion and Realism. Immersion is achieved with a plausible design that lets users suspend their disbelief. In many cases, this works *against* the goal of Realism and is also generally much easier to achieve.

User Experience: Haptic sensations should be easy for users to interpret, and they should naturally complement the scenario and content. Haptics optimized for user experience should have a shallow learning curve. Haptic sensations requiring a steep learning curve, such as complex sets of haptic patterns, should be avoided, especially for non-expert users.

Skills Transfer: Haptic feedback should generate net positive training metrics, including increased memorability and reduced error rates (criterion validity [23]). Providing feedback before or during interactions is typically more successful than negative feedback after user interaction. However, this feedback should be driven by the goal of refining a user's mental model.

A good test for the efficacy of haptics is to perform A/B testing between haptic and non-haptic conditions at key development milestones.

The most thoroughly researched domain for haptic skill transfer is surgical training. Further information can be found in [23].

Usability/Accessibility: A key difference between haptics and audio/visual technologies is that users typically wear haptic devices. This raises issues related to usability and accessibility. XR Haptics experiences should be accessible to various sized users and with varying physical affordance requirements. It is important to consider user size variation and the complexity and calibration of devices that need to be fitted to individual users. This will also drive considerations around throughput and other overhead associated with the end-to-end XR Haptics scenario.

Safety: Devices that generate forces on users need to be engineered so that they cannot cause injury due to inadvertent use. If possible, users and practitioners should be empowered to reduce/eliminate the haptic feedback to allow individual sensitivity and tolerance. Heavy use of haptic feedback devices for specific tasks also has the potential for repetitive strain injury (RSI). The workflow of the haptic application must incorporate sufficient breaks and rest periods to prevent users from developing RSI injuries.

Consistency: To use haptics to generate precise and measurable outcomes, the haptic feedback must have maximum consistency from both use-to-use and installation-to-installation. This is related to usability but is distinct in the sense that poorly fitting or challenging to calibrate devices are unlikely to be used correctly in an industrial scenario, where time is of the essence. For body-mounted feedback, careful consideration should be paid to fit consistency, including having multiple sizes or specific criteria for appropriate body size considerations.

Expressivity: Haptic devices need to provide a sufficiently perceptible range of experience so that users can correctly interpret the stimuli. Basic body-mounted error feedback may only need on/off expressivity (e.g., 1-bit), whereas haptic feedback based on a quantity of error measure may require many more bits. Also, note that the human body has a significant sensitivity variance. For example, fingers are much more sensitive than the back, and hairy skin is less sensitive than hairless skin. Device selection should be carefully guided by this consideration since low-fidelity haptics may have a negative impact on XR Haptics effectiveness, and high-fidelity haptics may be costly and unnecessary for a specific scenario.

Robustness: A haptic system should be reliable, particularly if the operator of the final system is non-technical or is an end-user. The throughput of a haptic system should be carefully considered based on the target use case. For example, sales and marketing consumer use cases tend to be used primarily by non-experts who need more reliability than high-end virtual prototyping systems.

Tip From the Expert:

When evaluating haptic devices for an XR scenario, be sure to spend time using the device when it is not actively generating haptic feedback to assess its transparency.

Chris Ullrich, Immersion Corp.

Transparency: When selecting hardware, transparency is one of the most important considerations. Transparency measures the baseline or background haptic sensation created by the physical system when it is unpowered or switched off. A device with many mechanical components, gearboxes, or other elements that have passive resistance will not feel transparent, even when no haptic effect is played. The result of low transparency is increased or rapid user fatigue, low realism, and a lowered potential for skill transfer. In the ideal case, a haptic device would convey no haptic sensation at all when switched off, yet be capable of high-force stiffness simulations. Practically, these two qualities are not simultaneously possible to achieve in the same device.

5.2. Actions Required

A best practice for sorting through the available haptic goals is to run a design process to get stakeholders to identify and prioritize them. If the XR scenario is already well articulated in a storyboard or wireframe, the stakeholders can actively review and debate each section of the experience and determine the role and objectives for haptics. This can be accomplished using dot voting or some other similar design technique. If the XR scenario is not well articulated, then a less structured approach, such as clustering, can be employed to identify the project's most essential haptic objectives. Stakeholders must assess the expectations relative to haptics in the scenario realistically. Having at least one haptic expert facilitate this action can dramatically improve the outcomes.

5.3. Key decision to make

During this step, it is essential to prioritize the haptic goals in collaboration with a representative selection of the project's users or customers. At a minimum, the implementor and the customer should align on the top 3 goals for the haptic component of the experience. Selecting only a single goal is not recommended. This may create unrealistic expectations and ultimately undermine the addition of haptics to an XR scenario. The top 3 goals can then be used to evaluate potential hardware solutions systematically.

5.4. Key Considerations

Compatibility of the objectives: It is unlikely that all goals can be met perfectly with a single system design. For example, highly realistic haptics tend to require complex and time-consuming interfaces, reducing portability and accessibility.

A first-level understanding: Haptic devices range from a few dollars to a few hundred thousand dollars. The program team should have a rough notion of what the budget will support as they evaluate the priority of the haptic objectives.

5.5. Deliverables

The deliverable of this phase is a document that defines the goals of the haptic information within the virtual experience and a level of importance of each objective; "haptic goals". These definitions will be used in the subsequent design steps, from the implementation to the users' evaluations, as metrics to validate if the goals are met. Chapter 11 will describe a template model, including an example of every deliverable discussed.

6. Haptic Interaction Design

6.1. Overview

Once the goals have been identified and prioritized, the next step is to identify the specific interactions that will benefit from the addition or use of haptic feedback. Each goal should have one or more associated interactions that achieve the objective. This chapter provides an overview of common haptic interactions, followed by the association of these interactions with the haptic objectives from Chapter 5.

Note that the haptic interactions proposed in this chapter are not exclusive and not presented in a particular order. Some authors have attempted to create taxonomies of interactions as a means to enable discussion of interaction ontologies. For practical purposes, these organizing principles do not greatly inform the implementation of haptics in an XR environment.

6.2. Haptic Interactions

Haptic interactions can be categorized as shown in the following illustrations. Implementors should familiarize themselves with them and use them during the design phase.

Name: Ambient Effects

Location: Hand and Body

Perception of vibrational behaviors present in the environment, such as vibrations induced by a nearby car or machinery.

Example: A train approaching the player from behind.

XR Haptics: Implementation and design guidelines

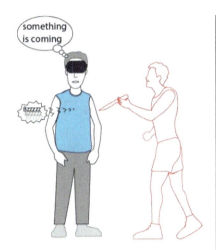

Name: Contextual awareness

Location: Hand and Body

Haptic feedback to have an enhanced understanding of the surrounding scenario.

Example: a vibrational pattern sent to the users to advise them of an upcoming danger.

Name: Dynamic Interactions

Location: Hands

Continuous haptic stimulation associated with the manipulation or interaction of the user with a virtual object.

Example: grab and move a lever, door, or slider.

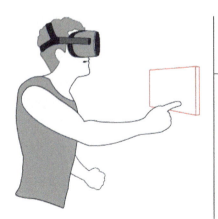

Name: Static Interactions

Location: Hands and body

Perception of contact with an object.

Example: Poking a tablet or a screen. Hitting an object while moving.

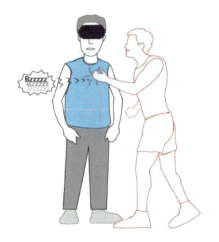

Name: Shocks

Location: Hand and Body

Perception of a shock on the body.
Example: A rapidly moving object hitting the body, such as a fist or club.

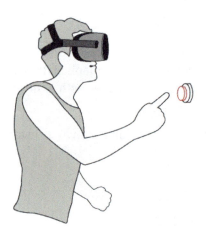

Name: Clicks and dynamic controls

Location: Hands

Touch perception of a real button or a dynamic switch. It is associated with a rapid transitory touch stimulus which confirms the execution of the digital interaction. It is useful to simulate the interaction with a virtual HMI screen in a virtual training scenario.
Example: Virtual button clicks.

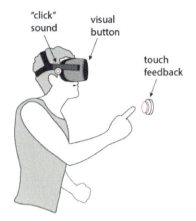

Name: Multisensory events

Location: Hands and body

Haptic Feedback in correlation with visual and audio cues to reinforce visual or audio objectives.
The other haptics interactions here listed might be designed as multisensory events. For example, a virtual button click might generate a haptic click effect and an audible click.

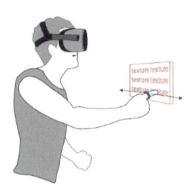

Name: Textures

Location: Hands

Perception of the surface texture of an object while sliding above its surface.
Example: Dragging a virtual hand over a virtual fabric surface, such as a couch or chair.

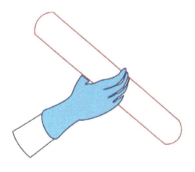

Name: Object Manipulation (Large)

Location: Hands

Perception of an object being manipulated.
Example: the manipulation of a ball held with the whole hand.
Note the difference between this and little object manipulation involves different interaction peripherals, the whole hand in this case, and the fingertip for small object manipulation.

Name: Object Manipulation (small**)**

Location: Hands

Perception of a small object being manipulated with the fingertips.
Example: the manipulation of a small key rotating in a keyhole.

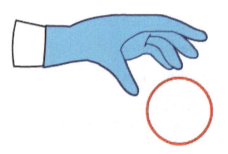

Name: Shape

Location: Hands

Perception of the shape of an object while held in hand or while manipulating it.
Example: the grasping and manipulation of a round ball.

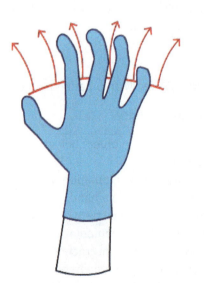

Name: Active objects (slow)

Location: Hands and body

Perception the modifying shape of an object.
Example: an inflating ball held in the hands.

Name: Active objects (fast)

Location: Hands and body

Perception of active objects or objects used as tools to interact with the environment.
Example: a beating heart or a drill in use.

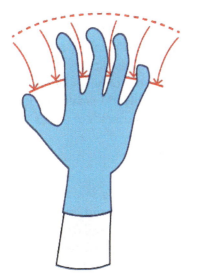

Name: Softness / Stiffness

Location: Hands and body

Perception of the compliance of an object. The property is usually experienced while poking or indenting the object with the human hand or finger.
Example: the softness of a pillow or a balloon.

6.3. Haptic Interaction-Objective Considerations

At this point, the implementor should have a prioritized set of XR objectives and a good understanding of possible haptic interactions. The following section connects the types of interactions that enable each haptic objective. The implementor should consider these items carefully to maximize the selection of a final set or prioritized list of desired interactions. The haptic objective listed above is here linked to the interactions just introduced.

Realism – hardware and software

Many of the listed haptic interactions contribute to creating a sense of realism for the user. Those that are interactive, such as clicks and dynamic effects, are particularly beneficial to generate a sense of belief by the user in the XR experience. Interactions with no real-world analogue (e.g., contextual awareness) reduce the sense of realism users feel unless they have a narrative reason to have this tactile experience.

Immersion – hardware and software

Immersion is the characteristic of enhancing the user's sense of virtual presence during the experience. Immersion is achieved by ensuring consistent haptic feedback integrated correctly with audio and visual feedback. Immersion and Realism are sometimes correlated and sometimes not. All the listed haptics interactions shared before can be used to enhance the Immersion objective if coupled with the right haptic technology and the correct design principles. Note that incongruous implementations, such as poorly synchronized haptic and audio feedback, have a significant negative impact on user immersion.

User experience – mostly software

Haptics optimized for user experience should have a minimal learning curve. Haptic sensations should be easy to interpret and naturally complement the scenario and overall user experience of the content. Haptic sensations that are nuanced and complex, such as haptic patterns, should be avoided, especially for non-expert users.

Haptic elements in an interaction such as a click or dynamic effect may enable users to execute tasks in the XR environment with less stress and less confusion. This UX value can be extremely important, especially when users are expected to spend significant time in the XR system. The specific implementation of the interactions has the greatest effect on user experience and should not be neglected during development.

For example, Clicks and dynamic events, contextual awareness, Interactions (Hand – Objects), and all the haptics interactions linked with manipulation and interactivity meaningfully contribute to the overall UX of the XR experience. Haptics interactions linked with realism like Shape or Softness usually do not bode well with user experience and tend to distract and confuse the users because they can be challenging to implement in a believable way.

Skill transfer – hardware and software

Skill transfer leverages interactions to create muscle memory or conceptual understanding of tasks. Interactions involving manipulating objects or tool usage contribute enormously to the transfer of skills from XR to the real world, provided that the haptic feedback does not create incorrect expectations from users. This is a situation where realism and immersion should be balanced to maximize skill transfer.

Usability / Accessibility – mostly hardware

The usability/accessibility objective is mainly correlated with the ergonomics of the Haptics hardware device choice and the high-level implementation of the scenario to respect the embodiment and immersion guidelines. Some key considerations here are workspace size and transparency of the hardware itself. From an interactivity perspective, making the interactions have consistent, easy-to-understand feedback greatly enhances user satisfaction and, therefore, the usability of the XR system.

Safety – mostly hardware

The safety objective is delivered mainly by choice of the hardware device. Implementers should pay attention to all the haptics interactions with an active and not reactive nature. For example, Object Active (Fast) interactions can be dangerous for the user if coupled with a strong force feedback device because the environment is actively applying a force on the users' articulations.

Consistency – mostly hardware

Many triggered interactions such as click and static have a high degree of consistency because they are essentially fixed effects rendered in an open-loop fashion. Interactions that depend on user motion, such as object manipulation and active objects, incorporate user tracking into the feedback loop. User tracking technologies are susceptible to noise, which may create inconsistent interactions from the user's perspective. If the user sensing component of the system is noisy, it is not recommended that dynamic or other closed-loop interactions be used.

Expressivity – hardware and software

Expressivity relates to interactions in two ways. For static interactions, more expressive haptics can create a greater range of sensations and reduce user fatigue relative to the XR experience. For dynamic interactions, expressivity enables a broader range of user-driven sensations, which can be very compelling for users and reinforce immersion and realism.

Robustness – hardware and software

Robustness is typically inversely correlated to other haptics objectives, such as Realism and Immersion. Extremely robust XR systems usually need to be simple and may not be compelling to users for very long. That said, interactions that rely on static feedback such as clicks or certain types of active devices are well suited to robustness. It is important to note that, from a hardware perspective, robustness typically reduces transparency due to increased redundancy in the haptic hardware itself.

Transparency – mostly hardware

Transparency can be a key objective for training since users must experience tactile feedback only when relevant to the training goal. In terms of interactions, transparency is typically dictated by the hardware choice. If specific interactions are essential to the success of the XR experience, especially in a training scenario, it is recommended that transparency be a key consideration in the hardware selection process.

6.4. Actions Required

Stakeholders and implementors should develop an interaction design document that characterizes the key interactions in the XR haptics environment for each goal identified in Chapter 5. This interaction design document should have a section associated with each haptic objective and identify the interactions required to enable that objective. This document's organization depends on how well the program team understands the entire XR scenario. For well-understood scenarios, it should be straightforward for the team to list the enabling objectives for each storyboard/user journey element. For less well-articulated scenarios, the team may need to ideate appropriate interactions in a more general setting.

Once complete, the interactions can be extracted and listed in a separate document, ideally prioritized. This will be highly beneficial during the hardware selection process.

6.5. Deliverables

The product team should now be able to list the haptic interactions that will need to be created to achieve each goal identified in the previous chapter. This document can be organized by the prioritized objectives and should identify the interaction design(s) in detail for each objective. A final list of prioritized interactions (based on the set in section 6.2) is the key connecting information between the larger haptic goals for the project and the specific technologies (hardware and software) that will be used to develop the XR scenario.

7. Hardware device choice

Tip From the Expert:

The way to ensure the success of the Haptics hardware design or choice is to include the final user as soon as possible in the loop.

Pf. Edward Colgate, Northwestern University, and Tanvas Co-Founder

7.1. Overview

The choice of haptic hardware to implement the XR scenario is the next step in the process and is likely one of the most critical. Haptic hardware can range from essentially free to hundreds of thousands of dollars, so it is essential that the choices made at this step optimize both the haptic objectives identified above and the budgetary constraints. The variance of cost, quality, and suitability of haptic hardware is one of the critical factors distinguishing it from audio and visual feedback technologies. In addition, many haptic hardware solutions have siloed development technology stacks and are not interoperable. For these reasons, the choice of haptic hardware is both the most complex and highest risk component of developing a haptic XR scenario.

This section contains a lot of detailed information intended to serve as an orientation for newcomers and as a reference for the product team. To aid in navigating this section, it is organized as follows: a taxonomy of the types of haptic feedback is presented in section 7.2, several unique considerations to haptic feedback are discussed in sections 7.3 and 7.4. Finally, the actions and key considerations close out the chapter starting in section 7.5.

7.2. Haptic Feedback Taxonomy

Before evaluating specific haptic device options, it is often worthwhile to consider the haptic modality that best fits the identified XR haptics objectives. XR haptic devices can be organized into a few key technology areas, within each of which there are similar costs/benefits.

For practical purposes, haptic technologies may be divided into four categories:

Tactile/Vibration – Feedback that stimulates users with a vibration sensation. This is, by far, the most common haptic feedback type as it is found in nearly all modern mobile phones and consumer gaming and VR peripherals.

Kinesthetic – Feedback that provides a sensation of force or resistance. This is relatively common in enterprise VR solutions but is uncommon in consumer products.

Tactile/Contact Spatial – Feedback that modulates sensation through surface contact, electrostimulation, or skin deformation, both planar and normal. This is uncommon but very effective for specific use cases that involve surface contact.

Tactile/Non-contact Spatial – Feedback that provides sensation without any physical contact with the user. Usually achieved using air pressure waves, this is uncommon but has tremendous potential for high-transparency haptic feedback.

Two other types of feedback, temperature feedback and electrostimulation, exist within the haptic domain, but they have not yet found an extensive commercial application. For this reason, they are not described in detail in this chapter.

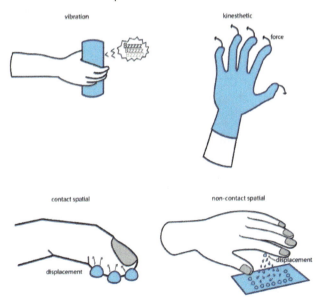

Figure 6: Representation of the 4 categories of haptic technologies available for haptics implementers

In [2], a basic taxonomy of commercially available haptic devices is presented. It is adapted here for reference.

Haptic feedback covers a wide range of possible stimuli but is broadly divided into two categories known as *tactile* and *kinesthetic*. Tactile feedback refers to sensations stimulating cutaneous receptors, such as vibration, friction, or deformation. Kinesthetic feedback refers to sensations that provide force sensations that can stimulate both mechanoreceptors and proprioceptors.

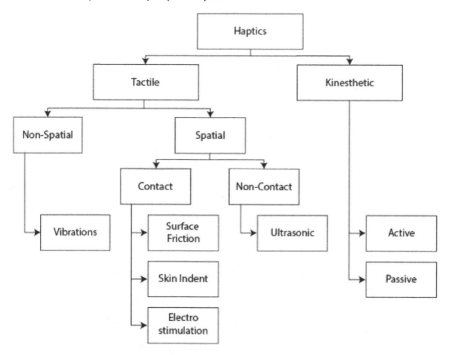

Figure 7: Taxonomy of haptic feedback hardware devices. Adapted from [3].

7.2.1. **Non-Spatial - Vibrotactile**

Vibrotactile haptics is the most widespread kind of haptics used in XR. Subtypes include ERMs (Eccentric Rotating Mass) actuators, which create a rumble-like sensation, the more expressive LRA (Linear Resonant Actuator), wide-band actuators like VCMs (Voice Coil Motors) and PZT (Piezoelectric Transducer) actuators. Devices equipped with wide-band actuators can deliver from a realistic click of a button to nuanced vibration patterns that can give rise to sensations of texture, which are particularly useful in XR scenarios. Vibration effects can be conceived as a kind of audio for the skin, assuming there is sufficient frequency response. In an enterprise XR application, vibration actuators may be used to simulate traditional UI elements (buttons and sliders) along with surface texture, as well as to substitute for more expensive and complex force feedback systems.

These technologies are usually coupled with finger and hand tracking technologies for VR glove implementation, but it does not stop here. There are devices using a network of vibrotactile actuators to stimulate body parts beyond the hands. These devices are often presented in the form of wearables for arms, legs, or a vest for the torso.

Examples of vibrotactile XR devices for hands are All VR controllers, Manus VR prime gloves, Bebop Sensors, where typical devices for body stimulation are haptic vests, like TactSuit x40 from bHaptics [3] or Skinetic [4] from Actronika.

7.2.2. Kinesthetic - Resistive Force Feedback

Resistive force feedback is used in haptic exoskeletons and gloves to impede the movement of the fingers in virtual reality. Resistive devices act as a brake for a finger or body motion. They are used to enable virtual manipulation and are effective at enhancing realism during manual interaction. Resistive force feedback devices are often based on electromechanical brakes, which increase friction on a sliding cable. The real-time modulation of this friction controls the resistive force experienced by the user. They are usually coupled with finger and hand tracking technologies.

Examples of resistive force feedback devices are Senseglove [5] and HaptX [6].

7.2.3. Kinesthetic - Active Force Feedback

Active force feedback is used in the haptic exoskeleton and handheld haptic devices to apply an active force on the joints. This type of feedback typically utilizes electromechanical motors that apply a force to a body part. This force simulates interaction with a virtual object or simulates a realistic interaction with a specialized haptic controller device like a rifle or flight stick.

These devices effectively simulate manipulation tasks and interaction with non-static virtual objects and simulate realistic behavior of physical interfaces.

Due to a more complex mechanical implementation, these devices tend to be more complex and expensive than resistive or vibrotactile devices.

Two examples of active force feedback devices are Dexmo [7] and Haption [8].

7.2.4. Contact Spatial - Surface Friction

Devices exist that can modulate the coefficient of friction on a touched surface. This is typically an uncommon requirement for XR use cases but can provide a unique value if appropriate. These technologies are perfectly suited for the rendering of contact textures. For XR, they would typically need to be used with other feedback sources, such as kinesthetic or vibrotactile.

Two examples of surface friction devices are Tanvas [9] and Hap2U [10].

7.2.5. Contact Spatial - Skin Indentation

Skin indentation devices are used to selectively compress the user's skin to create the sensation of interacting with objects. Skin indentation devices can be used to render vibrations pattern, textures, or light forces perception, giving them a large spectrum of expressivity.

Three examples of skin indentation devices are Haptx, Go Touch VR [11], Weart [12].

7.2.6. Contact Spatial - Electrostimulation

Electrostimulation creates a haptic sensation by applying a voltage to the outmost layer of the skin. A short burst of electrical current running through the skin generates a voltage potential difference, generating a sensation burst. The sensation is hard to relate to real-world tactile experiences because the voltage excites both mechanoreceptors for tactile sensation and nociceptors for pain sensation.

One novel use of electrostimulation is through the application of a voltage through a muscle to generate self-induced haptic sensations by causing a contraction of the muscles.

One example of electrostimulation is Teslasuit [13], both the glove and the vest.

7.2.7. Non-Contact Spatial - Ultrasound

Ultrasound devices generate haptic sensations by focusing acoustic pressure at a given location in space. At this location, acoustic radiation force is produced, which slightly indents the skin, generating a tactile effect that can be modulated to produce various sensations. Ultrasound devices are suitable for generating light, spatially distributed sensations.

One unique quality of non-contact haptic devices is that they do not need to be worn on the body yet can generate haptics in a large volume.

Two examples of ultrasound devices are Ultraleap [14] and Emerge [15].

Tip From the Expert:

A good [XR] experience depends on the affordance of spontaneous interaction. It should not require training nor habituation and, above all, should not restrict movement.

Prof. Vincent Hayward, Sorbonne University, and Actronika Co-Founder

---

From the perspective of a technical implementor, each of these technologies has pros and cons that primarily manifest in the types of interactions that they can support. The following table reviews some of the key interaction considerations that should be evaluated during device selection.

7.3. Technology – Interaction considerations

Each of the haptic modalities identified in the previous section has ideal interactions in the XR environment for which they provide the correct enabling hardware. This section reviews some considerations related to the haptic hardware modality and the XR interaction requirements.

Vibrations

Vibration technologies are low-cost, easy to use, and can facilitate various interactions. In general, vibration feedback should not be used continuously but rather for transient effects. This makes vibration particularly useful for clicks and dynamic controls, shocks, active objects (fast). Vibration can also notify the user of state changes, such as when grabbing a virtual object. In some cases, vibration is also suitable for surface texture rendering, although wide bandwidth (HD) (PZT, VCM) type components are needed to achieve good results. Vibration technologies should not be used for continuous feedback, such as object manipulation and stiffness, active objects (slow) except to display transient aspects of these interactions and generally in conjunction with other haptic technologies.

Kinesthetic

Kinesthetic technologies are typically the most expensive type of haptic device and the most difficult to control. That said, kinesthetic feedback is extremely compelling for slow, continuous types of interactions, including object manipulation, stiffness simulation, and object interactions. It is also possible to render high-quality textures with kinesthetic displays, but only if the control signal has a high sample rate (i.e., these displays will typically require high-resolution assets during haptic content creation). Many low-cost kinesthetic displays use brakes instead of actuators, which makes control easier but also makes them unable to render textures.

Active kinesthetic technologies can duplicate nearly any sensation that vibration feedback can generate, but only if they have sufficient bandwidth and actuator authority across the workspace. Due to the cost of achieving this, it is quite common to use a mixture of kinesthetic and vibrotactile feedback in a practical XR system.

Contact Spatial

Contact spatial displays normally consist of a 2D surface capable of generating variable friction, electrostimulation, localized sensations on a surface, or both. Perhaps the best example of an XR haptics device for contact spatial display is the HaptX DK2 gloves, which use pneumatic arrays to modulate surface texture on the fingertip.

Other contact spatial technologies are not widely deployed in XR Haptics use cases because of their relative inflexibility in spatial location. These devices are typically best-in-class for on-surface texture rendering and could be used in mixed reality use cases, where a physical substrate can have variable texture overlaid on its surface. Another consideration for applying contact spatial technologies is the cost consideration. The denser the pixel density of this technology, the higher the price of such a device will be. Devices that only have one "pixel" density per finger are priced like force-feedback devices. However, when the pixel density increases, the cost rises quite quickly.

Non-Contact Spatial

Non-Contact Spatial technologies can generate light pressure feedback on the skin from a distance without requiring users to touch, hold, or wear a specific apparatus physically. Modulating the position and intensity of the pressure over time enables non-contact spatial technology to produce vibrotactile-like stimuli on the palm and spatiotemporal patterns.

Non-contact spatial technologies are well adapted to convey continuous or short-time feedback, such as ambient effect, dynamic controls, contextual awareness, light interactions, and active objects.

However, non-contact spatial technologies have a limited dynamic range compared to other haptic technologies, making them less adapted to convey transient feedback or texture information. As such, object manipulation and shocks should be avoided with non-contact spatial technologies. In conjunction with hand-tracking, non-contact spatial haptic technologies can easily be deployed to XR haptics and haptify a wide interaction space. However, non-contact spatial technologies are limited. In conjunction with hand-tracking, non-contact spatial haptic technologies can easily be deployed to XR haptics and haptify a wide interaction space.

7.4. Related Haptic Solutions

In addition to the technologies described above, several 'hacks' can be used to create compelling experiences at a meager cost. Technically, neither approach in this section is a haptic technology, but they play a similar role to haptics in many real-world scenarios and may complement haptic technologies to dramatically improve the realism of an XR experience.

7.4.1. Passive Haptics Considerations

Haptic feedback that is integrated into a physical prop and is tracked in the simulation provides a meaningful and low-cost way to dramatically improve the realism of the haptics in an XR simulation. The tactile sensations provided by physical props are classified as passive haptics since they cannot be actuated or changed during the simulation experience.

An excellent example of a haptic physical prop is the StrikerVR Area Infinity [16] device which provides the physical form factor of a long gun with embedded haptic feedback that is typically used to simulate the tactile sensations of an actual gun. The passive haptic feedback, in this case, includes the surface texture, physical inertia arising from the mass of the weapon, and the feel of the hand grips. These passive elements are combined with various active elements to create an incredibly realistic haptic experience.

Contrast this with an XR implementation that tries to simulate the experience of holding a gun using an articulated exoskeleton, ex. Magtube from ProtubeVR [17]. In general, the simulated gun will have lower fidelity and incomplete haptic sensations than the prop because of cost and complexity constraints associated with the exoskeleton. For this reason, it is recommended to utilize passive haptics as much as possible in any XR project without unnecessarily limiting the range of possible experiences.

Passive haptic feedback provided by physical props such as weapons and tools is beneficial for certain XR use cases, particularly training and simulation. A specific physical interface can be instrumented with motion tracking and active haptic feedback (ex. section 15.2).

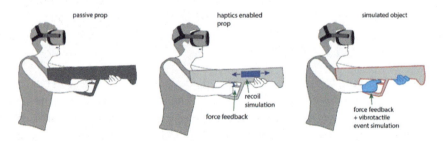

Figure 8: Passive Haptics, haptics enabled prop, and fully simulated interactable objects. For XR Haptics, passive props enable the implementation of believable scenarios at a meager cost.

Another class of passive haptics opportunity is projected UI over physical objects in XR. A simple example is the projection of a keyboard over a table and the leverage of the tactile feedback of the real object while interacting to generate a believable user experience. Because physical objects have extremely realistic and robust tactile properties, using projected visual and tactile elements in a mixed reality setting can provide high realism at a low cost. A key challenge of this approach is ensuring a high level of registration/calibration between the physical object and the projected visual and haptic content.

7.4.2. Pseudo Haptics

Using pseudo-haptic effects is a somewhat related and low-cost way to achieve haptic feedback. Pseudo-haptics is a type of haptic sensation generated by a discrepancy between the user's physical proprioception and the rendered avatar configuration. It is one form of multisensory illusion. More details on multisensory illusions are available in Chapter 10.

Tip From the Expert:

A lot of recent studies in pseudo-haptics show how nice it is and how visual modifications can give rise to haptic sensation.

Dr. Allison Okamura, Stanford University

For example, when a user grasps a virtual object, in most cases, the virtual graphical representation of the user's fingers will not be constrained to lie only on the surface of the object. While kinesthetic feedback would be useful to enforce this constraint, it is not typically reliable enough. A more immersive and realistic approach is to constrain the virtual graphical hand to lie only on the surface of a grasped object, as shown in Figure 9. Though this adds complexity to the hand tracking and visual rendering of the scene, the user will more readily suspend disbelief in the virtual environment and experience pseudo-haptic feedback based on the discrepancy between the virtual hand (visible) and their physical hand (not visible if using an HMD). [31]

Related pseudo-haptic techniques include redirected walking [18] and its analog for arms [19].

Because pseudo-haptic techniques do not rely on any physical haptic device, they can be highly cost-effective for achieving realism and immersive haptic objectives.

Note that there is added simulation complexity. Adding additional haptic cues above and beyond the pseudo-haptics significantly impacts realism and immersion in the virtual environment, similar to the gains obtained by adding haptic feedback to a passive haptic prop.

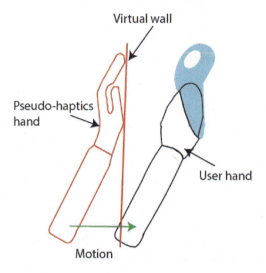

Figure 9: Pseudo-haptics representation

There are, of course, limits to the quality and magnitude of sensation that can be produced with pseudo-haptics. These limitations are essential to understand early in a project so that active haptic feedback technologies can be included as needed.

Several technologies that can facilitate pseudo-haptic sensations are listed in the references [20] [21] [22].

7.4.3. Self-Referenced Haptics

Another related haptic consideration is self-referenced haptics. This refers to the user's haptic sensations on their own body, such as a clenched fist or pinch grasp. For fully immersive XR use cases, it is important to remember that self-referenced haptics can expose poor sensing calibration since the user's virtual pose may not match their physical pose in space. Because they are generating self-referenced haptic sensation, this discrepancy may cause a loss of presence.

Self-referenced haptics can also be exploited to create compelling virtual interfaces. Consider a virtual UI that is associated with a user's forearm. By combining the soft compliance of the physical arm with a sharp transient haptic effect, it is possible to create a compelling and useful UI confirmation sensation. An example of self-referenced haptics is included in [23].

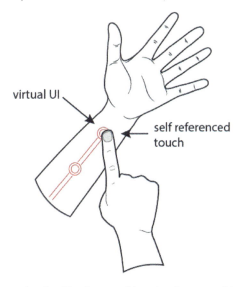

Figure 10: Example of self-referenced haptics for wearable UI interaction

7.5. Actions Required

The development team should carefully weigh the desired haptic interactions listed and prioritized in the previous section against available technologies. Cost is a key consideration for haptic systems and will likely be used as a gating criterion for the final system design. Candidate haptic solution vendors should be engaged and asked to provide details on the feasibility of implementing the prioritized interactions as well as the solution cost, including integration development costs, sensing dependencies, and any other required infrastructure.

As a general principle, including the final users within the hardware device choice process is of paramount importance, but it is not always possible to perform comprehensive user studies due to limited time or budget. This document and the following information are meant to support implementers in this choice and obtain a satisfactory result without a complete user test.

In most cases, there will not be an obvious choice of a specific haptics device or technology to meet the use cases objectives, and different devices meet the expectations for specific content.

> **Example:** A Virtual Training scenario including virtual HMI interactions with a Numerical Control Machine where the Haptics objective is to enhance the accessibility of the content; in this case, Optical Hand Tracking, Haptics Gloves equipped with vibrotactile actuators and Exoskeletons meets the haptics objectives, and the hardware choice falls on budget/implementation cost / global user experience desired for the content. In this case, it may make sense to start with the lower cost and lower risk vibrotactile hardware. Once this is working, it can be evaluated for deficiencies relative to the overall simulator requirements. It is quite common that force-feedback exoskeletons represent too much risk, complexity, and cost for XR experiences.

The development team should refer to the application area and use case guidance for recommended best practices for overlapping or similar application areas.

7.6. Deliverables

The deliverable from this section is the haptics hardware and technology choice to be included in the XR scenario. As stated above, this choice is critical due to the considerable overhead of switching haptic technology at this project stage. The choice of interactions in Chapter 6 should be listed against the capabilities of the selected hardware to ensure that the maximum set of interactions is supportable. Ultimately, this decision comes down to the cost of both the hardware and the integration effort to utilize the hardware. It is recommended that the program team revisit the hardware selection after considering the integration complexity described in subsequent sections.

7.7. Key Considerations

It is usually difficult, costly, or impossible to migrate an XR haptics scenario that has been developed for one category of haptic hardware to a different category. This is because each category's software and control technologies are unique and not standardized across these categories. It is possible to migrate or add support for other hardware products within the same category to an existing simulation.

For these reasons, the choice of technology category should be very carefully weighed against the haptic objectives as making a change after this point will likely be prohibitively expensive and/or result in significant re-engineering for the project. Two important considerations have to do with hardware and haptic layering.

As for hardware, the question is whether the application is to be designed to run on a specific set of devices or to be able to run on different hardware configurations. Depending on the importance of each desired virtual experience haptic objective, the hardware on which the experience is run may meet, or not, the objectives.

As for haptic layering, if we consider different haptic technologies to address a specific objective, it is also essential to consider the underlying complexity added to the implementation and the potential to realize the overall experience.

Tip From the Expert:

Don't forget that vibrations are underutilized, they are not just a buzz. Often, you do not need force feedback to achieve a determined design objective if you have a wideband vibrotactile actuator.

Dave Birnbaum, Creative Foresight

---

8. Haptic Development Environment

Once a haptic hardware platform is selected, it is necessary to plan and implement the integration of the haptic technology into the simulation. This chapter discusses some key considerations related to the development environment used for the entire simulation and how haptics can be most effectively integrated.

Tip From the Expert:

Minimize the amount of time between a haptics design change and the ability to feel it. Invest the time to have a tight feedback loop during haptics design, and design iteratively.

Dave Birnbaum, Creative Foresight

8.1. XR Development Frameworks

Most XR developers will take advantage of an existing simulation framework/game engine rather than develop the entire experience from the ground up. Both Unity and Unreal Engine provide extremely robust and extensible XR development environments, and both frameworks have some support for haptic feedback. However, if these systems are not appropriate, several lower-level frameworks and SDKs can still accelerate development.

8.1.1. Unity

Due to its extensive asset store and rich set of capabilities, many projects choose Unity as the framework for developing XR scenarios. Unity consists of a visual editing environment along with a game engine runtime and can generate executable binary applications for more than 20 popular computing platforms, including all major VR/AR/MR devices.

Unity provides a "HapticCapabilities" report about the hardware device capabilities. This functionality is restricted to multichannel vibration devices and offers only a very low-level interface, but it does support a mix of static and dynamic effects and individual channel addressability [24]. The only effect types that are supported are raw buffers and impulse. However, these can be combined with an audio engine or other audio systems to render high-quality vibration effects.

Due to platform popularity, many haptic hardware vendors provide plugins that enable easy integration of their devices into Unity. During the hardware selection process, the availability and scope of plugins to facilitate development should be considered in addition to the hardware cost.

8.1.2. Unreal

Unreal Engine is a competitive product to Unity and offers somewhat similar functionality. Unreal Engine is more commonly used for large-budget game productions, but it is more than capable of driving high fidelity haptic XR experiences.

Unreal Engine 4 provides both a C++ API and a blueprint controller for haptics (see Figure 11).

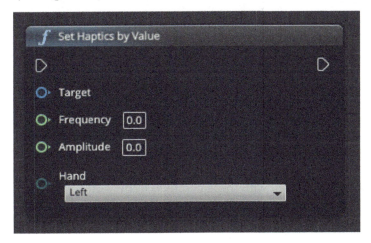

Figure 11: Unreal blueprint controller for haptics

The blueprint controller and API allow for fine-grained control of frequency and amplitude and targeting of the haptics to individual hands. This should be considered a low-level API and only suitable for use as a building block to create more complex and engaging effects. In some cases, this additional functionality is provided by hardware vendors, but in many cases, it must be developed as part of the project. Like Unity, Unreal Engine allows an audio engine to be used as a source for the buffers, enabling rich haptic experiences.

As with Unity, hardware vendors may provide plugins to enable rapid integration of haptic devices into the Unreal framework. If Unreal is chosen as the project platform, the integration cost of these plugins should be considered in addition to the hardware cost.

8.1.3. OpenXR

OpenXR™ is an emerging API standard developed by Khronos Group [25]. This standard is well suited for HMD-based XR applications as it enables programmatic, cross-platform immersive experiences. As with Unity and Unreal Engine, OpenXR provides support primarily for vibration hardware. This functionality is exposed using the XrHapticVibration structure, consisting of a dynamic list of duration, frequency, and amplitude values. This can provide control of frequency and amplitude but does not offer fine-grained buffer functionality for audio-like haptic waveforms. This limits the API to rudimentary haptic sensations, although it is always possible to directly generate haptic effects using device-specific APIs alongside OpenXR.

OpenXR supports vendor extensions, and it is reasonable to expect that haptic-specific extensions may be proposed and ratified by the Khronos OpenXR working group in the future.

In all cases, most XR haptic developers will need to use 3rd party extensions or develop custom device interface code to access the full functionality of any haptic device used with these common frameworks.

8.1.4. Device-Specific SDKs and Frameworks

In addition to the generic solutions described above, haptic device vendors may provide additional device-specific SDKs or frameworks. Oculus provided a framework in the past but migrated to OpenXR in 2021. Another example is the CoreHaptics™ framework for iOS, which does not have a counterpart on any other platform but can be used for mobile AR scenario development. When faced with the choice of a device-specific haptics development environment, the development team should consider future needs to support additional or different haptic devices. If the haptic hardware is unlikely to change, a device-specific SDK or framework typically offers the most efficient and cost-effective integration path. If the hardware may change, or there is a need to support multiple endpoints, then device-specific solutions may not be optimal long-term.

8.1.5. Generic Haptic SDKs and Frameworks

Generic haptics creation tools like those offered by Interhaptics [26] and Audiokinetic [27] enable the creation of haptic assets compatible with multiple platforms. The advantage of this approach is the ability to reuse the haptic assets and access specialized support for haptics development and design.

As of 2022, the haptic device/SDK ecosystem has been somewhat complex and fragmented, but there are several active standardization efforts (see Figure 12). One of the most important is the standardization of haptic asset encoding, which should simplify creating generic haptic assets and content and lower the cost of implementing haptic.

One example of a generic haptic tool that developers can use is Interhaptics, which is also associated with the draft MPEG standard for coded representation of haptic signals. Using a tool such as this can enable future-proof haptic systems by separating the designed content from the device-specific signals.

Generic haptic tools usually require a paid license and a development environment for the targeted platform.

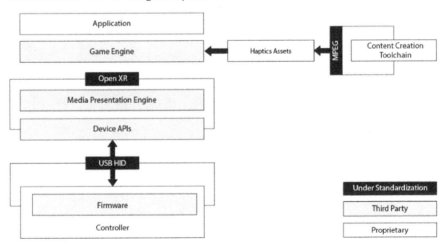

Figure 12: Exemplary diagram of a MPEG compliant haptics technology stack for a PC/XR configuration.

8.2. Haptic Effect Taxonomy

This section provides concrete examples of haptic effect creation for common XR use cases. In most cases, haptic assets are designed for a specific haptic type or need to be optimized for the hardware. There is currently no platform-independent haptic effect coding system, although, as described in the previous section, careful selection of middleware can maximize the range of technologies that can be supported. The runtime rendering system can also significantly impact the quality and value of the haptics. Effect-specific rendering considerations are discussed in this section, and general system rendering considerations in the next section.

For XR experiences, haptic effects can be broadly thought of as having three distinctive properties:

1. Static vs. dynamic
2. Vibrotactile vs. kinesthetic
3. Contact vs non-contact

8.2.1. Static vs. Dynamic

Static effects are haptic signals that do not change while rendering. They usually are one-dimensional signals but can also be multi-dimensional (e.g., surface texture). They are:

1. Usually easily authored with audio workflows and can leverage audio containers such as WAV.
2. Less engaging for users since the effects are always the same and usually repetitive.
3. Can often use simple API calls to trigger static and audio effects efficiently.

Dynamic effects are haptic signals that change as a function of the simulation state, including during user interaction with virtual objects. They are:

1. Described with function curves or algorithmically.
2. Provide a higher level of engagement for users as the effects are responsive to the user.

3. Require the developer to manage the rendering as part of the simulation loop.
4. Synthesized during the simulation based on surface texture, mechanical properties, or other simulation data.

Kinesthetic effects are generally dynamic and rely on high loop-rate feedback to calculate and determine the force applied at millisecond time scales.

8.2.2. Vibration Effects – static and dynamic effects

8.2.2.1. Transient - static effect

A common type of static effect is the transient vibration which consists of a short, high amplitude signal and is commonly used for mechanical feeling 'click' sensation. Although it does not appear in the latest models, the iPhone haptic home button was a high-quality transient vibration effect that felt like a real mechanical button. This effect is often provided only as a driver IC or firmware-level effect because of the extremely high rendering sample rates required (typically >5kHz) to create a sharp click sensation. This also has the advantage of very low latency. Since buttons are normally static and have a brief duration, it is acceptable to have limited software control over the effect itself during playback. Transient effects are often authored by specifying only the sharpness of the sensation and sometimes the amplitude.

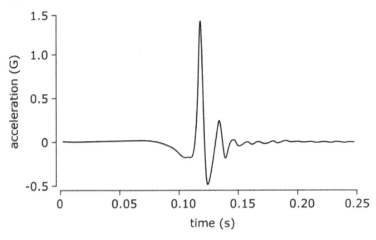

Figure 13: Typical sharp click effect used to simulate mechanical switches

8.2.2.2. Descriptive Effects – static effect

Descriptive effects are haptic effects represented in a signal-independent way, similar to MIDI for musical instruments, where only the notes and other 'descriptive' characteristics of the note are encoded. Apple's AHAP effect files (CoreHaptics) use a JSON-based descriptive effect language to enable expressive effect creation without needing to provide a motor signal. Some examples are shown below:

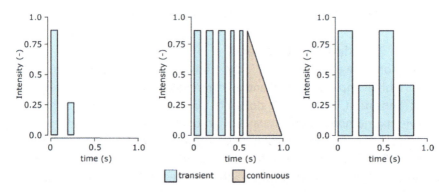

Figure 14: AHAP representation of haptics effects

Timed effects can be authored in a JSON-like format like AHAP or other XML-based human-readable formats.

Like MIDI, descriptive haptic effects need a preprocessor or runtime synthesizer to convert the description into a signal suitable for the available haptic hardware. This is commonly called a haptic player, and commercial solutions offer a variety of features, including mixing, audio synchronization, and real-time modulation of parameters.

8.2.2.3. High-Definition Vibration Effects - static or dynamic effects

The DualSense game controller created for Sony's PlayStation 5 gaming console includes both HD (high bandwidth) haptics and active force feedback on the trigger controls. Both capabilities are sophisticated and have a lot of potential for game and XR experiences. It is typical for audio assets and workflows to generate effects like these, particularly for HD vibration actuators.

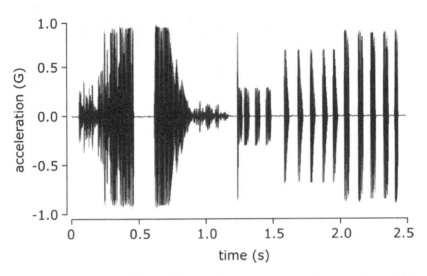

Figure 15: An explosion followed by gunshots generated with an audio workflow

There are haptic-specific rendering engines such as the Interhaptics engine that can provide a real-time synthesis of vibrotactile data [26]. Other audio-based engines, such as fmod [28] or Wwise [27], can also generate HD vibration effects in the form of audio PCM data, typically at audio sample rates of 8 kHz and above. These tools can also synthesize the PCM data dynamically using a model constructed by developers so that the actual experienced sensation is a function of user interaction or input. For example, a virtual surface could have velocity-dependent vibrotactile roughness generated by an audio engine. This technique is widely used for HD haptic devices, such as Dual Sense game controllers, since it both leverages existing audio rendering functionality and provides dynamic sensations that are typically more engaging to users.

8.2.2.4. Synthesized Vibration Effects – dynamic effects

Dynamic vibrotactile rendering techniques are functional effect descriptions that utilize a synthesis engine similar to descriptive effects but rely on external state variables, such as velocity or position, to modify an effect during playback. One example of a dynamic vibrotactile signal is a surface texture generated using the contact position on the surface of an object, Figure 16.

Figure 16: Example of dynamic vibrotactile to render in plane texture

Dynamic vibrotactile effects require a rendering engine to be implemented in the application's runtime to update the rate of rendering of a vibrotactile signal. The implementation of such a rendering engine can consume considerable resources depending on the complexity of the desired stimulus. As mentioned in the previous section, existing haptics or audio rendering engines such as Interhaptics engine, Wwise and fMod perform this exact task for dynamic audio synthesis and specific HD vibration devices and may be a good choice. These engines can be used at runtime with simulation state variables to synthesize rich and engaging haptic effects.

Special attention should be paid to the quality of the input signals for dynamic vibration effects. For example, if one of the parameters of a dynamic vibrotactile signal is mapped to a function of the user's position, the acquisition rate and precision of the position sensing system will introduce noise and delay into the effect synthesis. In the case of haptic texture based on spatial position, the dynamic vibrotactile signal must be modulated according to velocity. Basic techniques such as low-pass filtering may introduce latency and cause a poor user experience. If this portion of the system is critical to the overall haptic quality, it is recommended that a haptic middleware rendering solution be adopted or an expert resource engaged to ensure robust haptic signals.

8.2.2.5. Encoding of vibrotactile data

The encoding of vibrotactile data carries considerable importance when considering dynamic vibrotactile implementation. A signal-based approach can generate unwanted quantization artifacts when resampled during playback or when rendered on different haptic hardware than it was authored on. Preference should go to a descriptive and algorithmic encoding of vibrotactile data as this consumes much less bandwidth and is less sensitive to rendering artifacts that will reduce the overall simulation UX.

8.2.3. Kinesthetic Effects – dynamic effects

Kinesthetic effects create a force sensation on the user's body. Generating these sensations is more like a robot control problem than an audio design problem; hence, these effects are typically synthesized at runtime. For XR applications, generating forces on the user is almost always a result of a collision between the user's avatar and the virtual environment. The most common approach is presented in this guide, but a knowledgeable and creative development team could create variants or completely new effects.

8.2.3.1. Resistive kinesthetic feedback – dynamic effects

When designing kinesthetic feedback, a collision detection engine capable of providing continuous contact information at a high rate is required. This typically implies a physics or simulation engine running at a high frame rate as well. For kinesthetic rendering, the frame rate variance has an outsized impact on the quality of the haptic experience. Maintaining a consistent frame rate is essential for high quality.

In a typical framework based on a collision engine, three functions are used to drive the haptic feedback:

On Collision Enter: The collider determines a point in 3D that represents the contact location between the user avatar object and the scenario object. For example, if the user is using a data glove to pick up a virtual rubber ball, the virtual finger phalange that first contacts the ball will need to record the contact point. Sometimes a transient static effect is played at this point to indicate initial contact with the user. See Figure 17(a).

On Collision Update: During sustained contact, the point where the avatar and the object are in contact may change and needs to be updated. If virtual object non-penetration is being simulated (for example, with a physics engine), then there is typically a discrepancy between the physically tracked position of the user and the virtual avatar. This discrepancy can be used to generate a force vector suitable for dynamic kinesthetic feedback. See Figure 17(b).

On Collision Exit: Once contact is released between the user's avatar and the virtual objects, the haptic force can be reset.

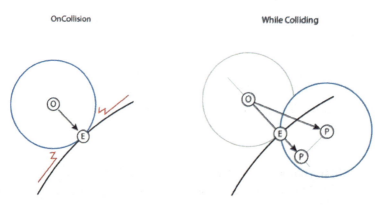

Figure 17: Representation of On collision Enter rendering strategy. A vector is calculated based on the haptic interface (o) collider relative to the feedback object collider €. In subsequent frames, the new position of the haptic interface position (P) is projected onto this Entry Vector (OE) to create t'e 'r'al' depth inside the material (P). A material script can be created based on this Entry Vector (x ax) and the amount of force resistance (y ax).

The actual calculation of the haptic signal during the collision update is determined by the desired interaction model in the simulation. A common choice is to create a function that depends on the penetration depth, such as the ones shown in Figure 18.

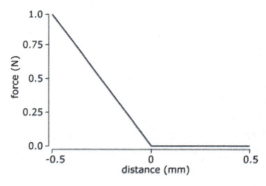

Figure 18: Force-distance function driving the kinesthetic feedback during collision.

For example, a soft, squishy object would start with a low force resistance and slowly ramp up to a high force value. While a breakable object will start with a high force resistance and drop to no force value after a certain penetration depth. The details of how these effect designs relate to the required interactions in the simulation are discussed in the next chapter.

Dynamic kinesthetic effects of this type depend on having high loop rates to calculate the penetration depth. Note that this calculation depends on the user's sensor latency and any collision detection processing latency. It is critical to minimize both this latency and its variance from frame to frame.

Device manufacturers usually release SDKs, allowing content implementation based on the devices, for example, the Senseglove SDK [5]. There is existing commercial software enabling the design and implementation of kinesthetic signals with different degrees of complexity. One example is Interhaptics SDK [26].

8.2.3.2. Active kinesthetic – dynamic effects

Another class of kinesthetic devices requires that all the processing for the haptics rendering is done on the device, while the host is not included in the control loop. This is typically done to provide a guaranteed minimum latency and low jitter for the haptic rendering. One example is the DualSense™ controller of the PlayStation 5 or the devices from StrikerVR [16]. Active kinesthetic devices are complex to control because they can become unstable. It is recommended that device-provided SDKs and associated effect descriptions be used as much as possible when deploying these types of devices.

8.2.4. **Non-Contact Spatial – static and dynamic effects**

Non-contact spatial are heavily dependent on the device manufacturer SDK and should refer to the provided documentation for implementers reported in the following reference [14].

8.3. Implementation considerations

Haptic feedback is fundamentally a closed-loop experience. As users move their hands or bodies, they will collide with other 3D objects. The body's movement is sensed proprioceptively and creates an expectation of feedback on the part of the user. If the immersive simulator has latency due to low frame rate, latency from sensors, slow haptic actuators, or any other latency source, this will create a discrepancy between user expectations and actual tactile sensation. This can create confusion and reduce the effectiveness of the XR haptics application and should be minimized as much as possible. A good rule of thumb is to keep round-trip latency from user position sensing to perceived haptic sensation less than 100 milliseconds. However, this number can be much lower for use cases with fast movement.

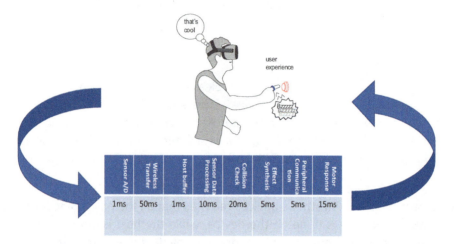

Figure 19: End-to-end latency for haptics is a key source of poor user experience. Here, sensor to feedback latency is greater than 100 milliseconds.

8.4. Encoding Standardization of Haptics Data Considerations

As the reader can appreciate, the effect encoding for haptics is variegated at best or dramatically fragmented at worst. Lately, the haptics ecosystem is gathering to find a suitable encoding standard for most of the presented use cases to facilitate the role of implementers and creators alike. The authors suggest following the evolution of the Haptics MPEG standardization process to select a reliable encoding standard for haptics effects. Industry bodies such as Haptics Industry Forum [1] provide a way to remain up to date and influence the development of haptic standards.

8.5. Actions Required

During this phase, the development team must align on a choice of development environment and selecting SDKs and frameworks. The objective is to facilitate the integration of the haptic hardware and support effect design and rapid iteration. Based on the selected or shortlisted haptic hardware, the development team should plan out the implementation strategy for the haptics from a software and experience design perspective. This phase should include the following decision components:

1. Simulation framework and haptic development environment
2. List of interactions and their haptic effect types
3. Plan for designing the listed haptic effect types
4. Collision detection and haptic player software architecture

8.6. Deliverables

At the end of this phase, the development team should have a design document and implementation plan for the software infrastructure needed to implement the entire simulation and, more specifically, the haptic portion. There should be a list of all haptic effects in the simulation and their associated creation or synthesis plan.

9. 3D Design Import, Haptics, and Interaction Design

9.1. Overview

This chapter deals with the haptic content portion of the XR environment. Beyond just determining when a haptic effect should be rendered, the association between the haptic content and the 3D scene, the interaction framework should be determined as this is the central runtime enabler of haptics.

9.2. 3D Design Import

Immersive simulation frameworks such as Unity and Unreal provide rich asset import functionality and have a robust ecosystem of asset import tooling. The critical consideration for geometry and environment design import is how interactivity will be enabled with the target geometry. In some cases, the visual geometry is too complex to enable consistent frame rate real-time collision detection required for haptic feedback. In these cases, a key activity is to generate high-performance haptic versions of the collision geometry. This geometry does not need to be rendered visually but is instead used for simulation.

The detail required for the collision geometry depends on the choice of interactions and their associated haptic feedback effects. The more dynamic or synthesized the haptics will be, the more detail is needed for collision geometry. For example, in a simulation where a user picks up a tool, the following table summarizes two possible geometry requirements:

Interaction	Haptic Effect	Geometry Requirement
The user's avatar snaps the object to hand when it is near. The user feels a thud sensation to alert them that a tool has been picked up.	Static vibrotactile	Basic bounding box of the tool
The user's avatar picks up the object with articulated fingers and can identify the correct grip location based on simulated surface texture.	Dynamic vibrotactile texture	High-resolution triangle mesh or voxel occupancy map.

For XR workflows in Unity, the Pixyz toolchain has been broadly deployed in aerospace and automotive markets. Similarly, Unreal Engine includes build-in functions to import complex 3D objects. Importing realistic models can be an essential component of the visual aspect of the XR experience, but it is likely to cause performance and usability issues for the haptic portion. There are also available tools and SDKs built for XR Haptics that address these issues with a pre-built framework [23].

9.3. Haptics-Interaction Framework

Haptic effects are usually generated in the simulation due to user interactions with the scene, although they can also be generated as ambient sensations in some situations. For our purposes, the following interaction framework provides a useful way to evaluate practical choices.

We will introduce the concept of Egocentric and Allocentric haptics to help navigate XR haptics interactivity. [32]

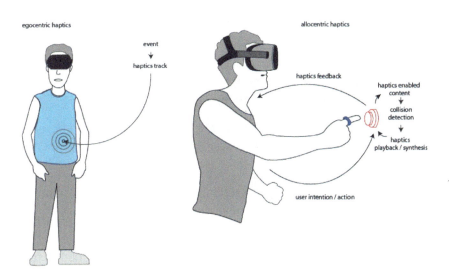

Figure 20: Egocentric haptics (left), the haptics track is played on the body following an event. Allocentric haptics (right) the haptics is described as a property of the content and executed in the function of the user intention or action.

9.3.1. Egocentric Haptics

Egocentric haptics refers to the scenario in which haptics play back on a specific target body part following a scene event. Once an event starts the haptic playback, the scenario carries on the haptics rendering on the defined endpoint until it is finished or interrupted by the scenario itself. The static haptic effect defined in section 8.2 is usually played through an Egocentric Haptics-interaction approach.

The haptic interactions listed in section 6.2 usually associated with an Egocentric Haptics interaction framework are listed below with a practical example.

> **Ambient effects:** One way to render a pattern of rain with a haptic suite is by defining a static pattern across a set of haptic actuators and trigger this when the user starts the simulation.
>
> **Contextual Awareness:** A motorcycle is approaching the player from behind. A vibration pattern is played on the user's chest to alert of the incoming danger.
>
> **Clicks and Dynamic controls:** One method of delivering clicks and dynamic controls is to create an in-scene event playing back the haptic track on the user actuator associated with a specific body part, for example, a controller.
>
> **Multisensory Events:** These are usually triggered static effects played in tight synchronization across multiple modalities (video/audio/haptic).
>
> **Shocks:** An incoming projectile can generate a playback on a pre-defined body part by the developer.

Egocentric haptics have many limitations when considering user interactivity and user experience consistency. Effects are defined in space, and a pre-scripted playback of haptics occurs based on user location or other simulation states. Allocentric haptics can typically be implemented in conjunction with an audio engine or with simple triggering API calls.

One of the key implementation considerations of an egocentric haptics approach is to ensure the synchronicity of the haptics track with the rest of the content once started by the event.

9.3.2. Allocentric Haptics

In allocentric haptics, haptics-interactions are defined within the scene, and the haptics sensations are generated as a function of the user's interactions with the scene. Allocentric haptics is especially useful for creating a sense of immersion and presence for users.

Let's take the previous haptics-examples and hypothesize an allocentric haptics implementation:

> **Ambient effects:** An allocentric method of rendering rain is to create falling colliders from the top part of the scene with playback on collision trigger with the collided body part.
>
> **Contextual Awareness:** A motorcycle has a large haptics collider (the active area surrounding the vehicle) moving with the in-game character. Once the motorcycle approaches the user, the user's avatar enters the motorcycle haptics collider, which triggers the haptic pattern playback.
>
> **Clicks and Dynamic controls:** The button is associated with the playback of a click once a certain displacement is met. The body part on which it is played is the one pressing the button.
>
> **Multisensory Events:** Similar to the previous example, multisensory events can be defined within the 3D scene and triggered by user intention or interaction. However, they may depend on more subtle ways of user interactions. For example, the tactile sensation and audible sound associated with rubbing sandpaper.
>
> **Shocks:** An incoming projectile colliding with a specific body part generates haptics feedback in the precise location where it collided.

Allocentric haptics brings many challenges for a robust implementation. The interaction loop described in the next section should work properly. A robust haptics playback/synthesis pipeline should be in place, as well as a solid synchronization mechanism between audio and video. Device SDKs [5] [14] [29] and commercial implementations [26] provide different degrees of allocentric haptics software stacks.

9.4. Interaction Loop

The detection and management of interactions is a governing component of the haptic feedback provided in an immersive simulation, see figure 19, right. The following considerations are paramount.

9.4.1. Collision detection

There will be a 3D scene description that consists of geometric representations suitable for detecting spatial overlap (collisions) between objects in the scene. In many non-haptic simulations, bounding boxes are sufficient for most simulation goals. In haptic simulations, bounding boxes only provide enough information to generate static haptic effects. This is because the collision function only provides a contact/no-contact Boolean at the collision start and finish. Dynamic haptic feedback requires a more sophisticated collision calculation:

1. **Proximity** – Often, a kinesthetic device will work best when the device's firmware is given a position-dependent effect. This can be thought of as a response function in the joint space of the haptic actuators. The haptic device will provide better performance and stability if the device has a small region around the non-zero force-feedback start point (in joint space). A collision detection algorithm that can report proximity will be able to provide this to the device.

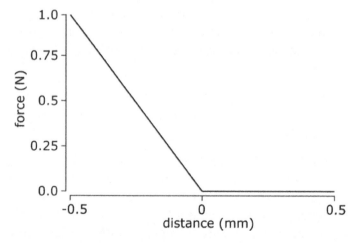

Figure 21: Force feedback as a function of distance from the surface of a 3D object. This function can enable high-performance haptic feedback if the calculated distance value is reasonably continuous.

2. **Contact tracking** - During sustained contact, the position of the contact point will typically modulate the haptic sensation (due to variance in surface texture, geometry, or other simulated material properties). A collision engine able to track the penetration or other on-surface contact region will enable a richer simulation of sustained contact.
3. **Penetration depth** – During manipulation, it is often useful to generate kinesthetic feedback as a function of real-virtual discrepancy (see 8.2.3.1). To calculate this discrepancy, it is necessary to calculate the penetration depth of the sensed user position vs. the virtual position. This calculation is straightforward for point-like/spherical avatars but is more complex for articulated avatars (e.g., a virtual hand). However, several good packages described in section **7.4.2** provide this information in real-time.

9.4.2. Internal Dynamics

Beyond basic geometric collision detection, it is also possible to have state-dependent virtual object functionality. A good example is a user pressing a virtual push-button. In addition to the geometry of the push button itself, the button has a single degree of freedom that the user can explore by pushing (with their avatar) along the primary button axis. A push button may have a kinesthetic and vibrotactile haptic response profile as it moves along its primary axis. The internal state of the button may trigger continuous and triggered static haptic sensations as it is manipulated.

The specifics of object state-dependent haptic effect as described in section 8.2 are unique to each simulation use case, and this is where there is an art to creating a rich, immersive experience.

Figure 22: Example of Internal dynamic interaction. The button touch feedback can be modulated in the function of the button status, position, velocity, geometry, etc.

9.4.3. Events (non-diegetic)

Haptic feedback may also be used for non-simulation events – sometimes referred to as non-diegetic effects – since they are 'outside the world' of the virtual environment. These effects are most often used by 3D UI that would like to provide tactile affordances and confirmation feedback on control interactions. For example, a pop-up menu may select simulation configurations or adjust simulation parameters from within the virtual environment. It is beneficial to provide haptic positional feedback to users to facilitate the easier spatial location of the 3D UI. This can be in the form of a transient vibration when the user is in proximity to the 3D UI, or ideally, it is some type of grounded kinesthetic feedback to help users locate and remain in the plane of the 3D UI elements.

Another possible use for non-diegetic feedback is to provide users with confirmation or error conditions related to the start or finish of task elements. For example, in a training use case, users may receive a specific haptic notification effect to indicate the beginning of a training sequence and a similar or distinct effect to indicate the completion of a task. A key consideration here is to ensure that the non-diegetic effects cannot be confused by users with other possibly similar feeling diegetic effects.

9.4.4. Haptics rendering and mixing

Haptic effects may be generated from multiple collisions, events, and ambient sources, and it is necessary to mix multiple simultaneous haptic effects for a single hardware actuation output channel. Some commercial haptic devices provide mixing functionality in their firmware, whereas others rely on the host to mix haptic effects. If available, it is recommended to utilize a 3rd party, perceptually aware, mixing solution since addition and normalization of haptic signals has sensitive dependencies on the device and human mechanoreceptor nonlinearities.

The mixing between different haptics effects usually follows the following strategies in function of achieving the objective.

> **Linear Mixing** - Effects are summed, and the resulting signal is delivered to the actuator. This strategy works well with wideband actuators, where the mixing stage is like the audio pipeline.
>
> **Priority Mixing** - some effects have priority over others. Higher priority effects cancel out or delay lower priority effects. One

example is core haptics from Apple's implementation of haptics. The transient effect cancels the continuous effects.

Non-Linear Mixing - The mixing between effects follows nonlinear functions. This is useful when dealing with amplitude-controlled actuators where a simple sum of effects generates saturation of the final signal, reducing the experience's quality.

A haptic mixer must also have a clipping and compression component since, for example, summing two full-strength effects would produce unexpected sensations depending on the relative phase of the two effects.

Tips From the Expert:
Build haptics as the representation of the perceptual objectives the interaction should deliver instead to the simple representation of the physics of the interaction.
Dr. Jess Hartcher, Facebook Reality Labs

Haptic feedback devices usually perform best when the final motor outputs operate with a very low temporal jitter. In an immersive virtual environment with a variable frame rate, dynamic haptic effects will generally create a poor user experience. A preferable approach is to have a haptic-specific processing thread that can operate at a high (100-1000Hz) frame rate and provide haptic commands with low jitter.

There are commercial solutions offering off-the-shelf haptics rendering processes, including mixing such as Interhaptics engine [26].

9.5. Actions Required

During this phase of the development process, the development team will need to identify and begin prototyping the 3D, audio, and haptic assets required for the simulation. Some of these assets may be static; some may be algorithmic (e.g., for dynamic haptics). The relationship between 3D geometry and haptic feedback cannot be overstated and should be prototyped as early as possible to determine the appropriate level of detail vs. the acceptable latency in the simulation. In many cases, having no haptics is preferable to having poorly synchronized haptics. As stated during the chapter, device SDKs and commercial solutions will help to alleviate the burden of building an in-house pipeline.

This activity is normally executed by a mix of software developers and asset designers working closely to find appropriate trade-offs to maximize the overall experience. A prototype experience with representative haptic hardware is needed to succeed in this step. Having easy-to-use tooling to enable rapid design iteration functionality early in the process is often the determining factor between an acceptable haptic integration and a delightful one.

9.6. Deliverables

The development team should have a list of interactions and their associated rendering simulation mechanic. The design team should have a list of required assets and begin creating, integrating, and testing these assets. Ideally, a rapid iteration activity is available to facilitate convergence on near-final assets as efficiently as possible. The interplay between interactions and assets should be functionally implemented at this step and suitable for stakeholder review and testing before assets and interactions are finalized.

10. Multimodal Integration

10.1. Overview

In the final chapter on implementation, we consider the role of multimodal integration in developing an XR simulation. Ensuring harmony between the sensory modalities of vision, hearing, and touch is the key to creating delightful, engaging, and convincing simulations. This is also challenging to achieve because of complex perceptual and technological interdependencies across assets, rendering, and interaction.

Almost any XR application is essentially multimodal since feedback within an interaction normally leverages cues from several senses. To maximize immersion and realism, haptic elements can be used to produce multimodal experiences. Haptic elements can take on different roles depending on the interaction objectives, as shown in Figure 23. Furthermore, design constraints must be observed to ensure the effectiveness of the multimodal experience. This chapter will briefly describe each role haptic element can take in multimodal interaction.

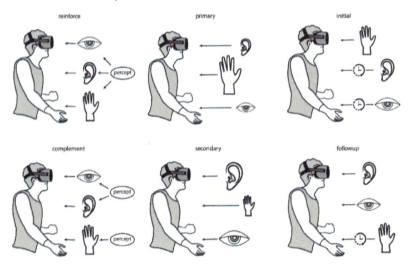

Figure 23: Roles a haptic element can take in a multimodal interaction [30].

Reinforce: haptics to reinforce the experience from the other senses with stimulus consistent with the meaning of audio and video modalities. This is one of the most robust implementations for extended reality experiences.

Complement: the role of haptics is to complement the perception with a secondary cue delivering a parallel communication channel between the scene and the user. This case, if poorly designed, can easily be interpreted by the user as inconsistent and generating discomfort. It is, however, powerful in specific cases to create a parallelized communication with the user.

Primary: haptics is the primary sensory modality for an experience. This is often the case in the real world.

Secondary: haptics works as a support of the other sensory modalities. This is the usual case in XR, where the tactile input is usually delivered by haptics technologies that render a reduced spectrum of the human tactile perception.

Initial: in this case, haptics is perceived before the other sensory modalities.

Tips From the Expert:

Consider building a library of interactions that are tunable in function of the multisensory experience you want to deliver in the XR environment.

Dr. Jess Hartcher, Facebook Reality Labs

Follow Up: in this case, haptics is perceived after other sensory modalities.

In the case of XR and for the purpose of the vast majority of the commercial implementation of haptics, implementers should follow these guidelines to achieve a harmonious multimodal experience for users:

Synchronicity: Timing between modalities is essential. Multimodal feedback that is not rendered simultaneously or the user will experience confusion and frustration as their brain attempts to integrate poorly synchronized signals. At a minimum, this will negatively affect the sense of immersion.

Congruency: Modalities should convey similar and complimentary information. Consider a bouncing virtual ball. If the visual bouncing motion implies a light soccer ball, but the audio conveys a heavy 'thud,' users will be confused, and the sense of realism will be negatively impacted.

Tips From the Expert:

Leverage illusions as much as possible to maximize the perceived realism of the interaction in a multisensory environment.

Dr. Jess Hartcher, Facebook Reality Labs

Modality predominancy: When presented with incongruent information, the brain will often trust one modality more than the other. Generally, visual is the predominant cue, followed by Audio and then by Haptics.

If a prototype simulation experience is not achieving its desired outcomes, careful tuning of the multimodal rendering can often improve performance in meaningful ways.

10.2. Implementation Guidelines

For this guide, implementers should consider that vision is the predominant modality in extended reality, and both audio and haptics should respect congruency and synchronicity with the visual stimulus.

This statement simply means that the user should experience the haptics feedback that the visual representation suggests and with similar timing.

10.2.1.   Congruency Guidelines

Congruency is the most complex aspect of multisensory integration for Haptics because the space for possible stimuli from haptics devices is limited compared to the expected experiences by the users. It often means that a common strategy is **not** to implement haptics to avoid inconsistencies in multisensory interactions. The challenge will be clarified in the following examples:

Example - manipulating a virtual object with a force feedback exoskeleton: If we implement a dynamic penetration effect while statically grasping a virtual object as described in section 8.2.3, the user should expect constant force feedback, even if they slightly move their fingers. In this interaction, the visual representation of the user's hand will not move if the user moves their fingers slightly, but the force may change meaningfully. If the haptics is modified without modifying the visual stimulus, the user will experience confusion, breaking the immersion. If the visual hand represented is static, the haptics should be static regardless of what the user's hand is doing.

Example - manipulating a virtual object with a fingertip-mounted vibrotactile actuator: if we select a wideband vibrotactile actuator mounted on the fingertip as a haptic device, it will be challenging to implement a congruent multimodal interaction while holding a virtual object steadily. The user expects continuous contact or force, and the vibrotactile actuator cannot deliver the desired experience. The best strategy, in this case, is to not implement haptics feedback continuously during contact and rely on the pseudo-haptics generated by non-penetration visual illusion and a transient haptic effect only during initial contact.

10.2.2.   Synchronicity guidelines

When implementing haptics and audio in a multisensory environment where vision is predominant, implementers should respect these maximum intermodal rendering delays:

Visual / Event Haptics	Visual / Contact Haptics	Visual / Acoustic
< 20 ms	< 40 ms	< 18 ms

The delay between Audio and Haptics is not important if both respect the maximum delay form Visual.

Further note that many haptic actuators take some time to achieve a threshold of perception level of tactile stimulation. This could result in poor synchronicity even though the simulation has very tight performance tolerances.

10.2.3.   Multimodal Implementation example[1]

Consider a simple example of simulating the multisensory experience during tapping on a virtual keyboard in VR.

How can we improve the user performance and experience by leveraging predominance, congruency, and synchronicity?

---

[1] Provided by Dr. Jess Hartcher from Facebook Reality Labs

Only visual: Change the color of the keys when the user approaches a key and change color again when they contact it. An 'Only visual' feedback implementation generates multiple unwanted collisions, making it impossible to type. Users get frustrated and confused by the visual feedback and ultimately have low performance.

Acoustic Feedback: Adding acoustic feedback on the collision increases the performance by decreasing the distance travelled by the finger because users can leverage the audio cue without visual attention to modulate their muscle motions. The overall result is significantly higher performance than visual only.

Haptics Feedback: Adding a slight transient vibration during contact reinforces the user's belief in the tangible reality of the virtual key. This does not provide performance gains over audio feedback but significantly improves user confidence and decreases cognitive loading.

In this example, it is possible to achieve the performance, immersive, and realism gains of each modality by combining them. However, these gains will not be additive if the multimodal implementation guidelines for synchronicity, congruency, and predominancy are not respected.

10.3. Actions Required

Multimodal integration should be viewed both as an interaction design activity and a simulation tuning and refinement activity. With a functional prototype, the development and design team should engage in user evaluation and tuning to adjust the balance of visual, audio, and haptic feedback modalities in the scene. This can include adding or removing feedback from any modality as well as adjusting the relative magnitude and timing of each modality until the perfect, delightful, and immersive simulation experience is achieved.

10.4. Deliverables

The activities related to multimodal integration should occur during every design-development-testing iteration. Ideally, multimodal integration leads to consistent, measurable improvements in the overall user experience.

11. Haptic design framework for XR

This chapter provides a worked example of the framework from chapters 4-10 that can be used as a template for implementing haptics into your XR scenario. We will explain the framework and take the reader on a step-by-step implementation with an example use-case.

The proposed framework corresponds with the chapters of the first part of this book. Detailed information can be found in chapters 4 to 10 to help you fill out the framework. The framework is meant to be used by project managers in collaboration with creative leads and programmers within the process of integrating haptics into your XR scenario.

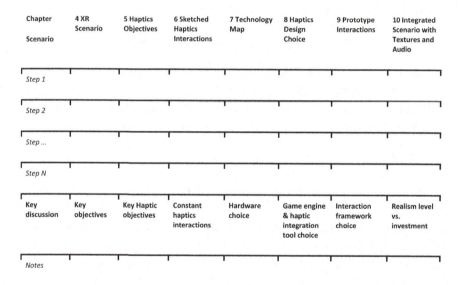

Chapter Scenario	4 XR Scenario	5 Haptics Objectives	6 Sketched Haptics Interactions	7 Technology Map	8 Haptics Design Choice	9 Prototype Interactions	10 Integrated Scenario with Textures and Audio
Step 1							
Step 2							
Step …							
Step N							
Key discussion	Key objectives	Key Haptic objectives	Constant haptics interactions	Hardware choice	Game engine & haptic integration tool choice	Interaction framework choice	Realism level vs. investment
Notes							

Table: Empty canvas to include haptics in an XR scenario

11.1. Example scenario

In the following sections, the completion of the framework is discussed for a relatable scenario for XR haptics training. To illustrate the implementation of this book's methodology, a worked example is provided in this section.

The scenario is as follows:

Build an immersive VR training simulator that can train the proper method for removing and replacing a car tire using a pneumatic, powered wheel gun (pneumatic driver) as is found in auto repair shops.

This trainer will be deployed and used to train mechanics by the manufacturer of the wheel gun.

Figure 24: Exemplary XR training scenario intended to train mechanics to replace a wheel using a pneumatic tire gun

The framework presented in the previous section is compiled for the selected use case in the next table.

XR Haptics: Implementation and design guidelines

Chapter Scenario	4 - XR Scenario	5 - Haptics Objectives	6 - Sketched Haptics Interactions	7 - Technology Map	8 - Haptics Design Choice	9 - Prototype Interactions	10 - Integrated Scenario with Textures and Audio
Step 1	Take the jack and place it under the right part of the car.	Immersion, Expressivity & Transparency	Object manipulation large (two-handed)	Vibrotactile & Resistive Force Feedback	On grab stiffness to the maximum for both hands. God hand pseudo haptics.	*Generic interaction design framework:* Ex. Interhaptics. Identify the interaction strategy which applies to them and drag and drop them on your scenario. Device dependent	Realistic sound when removing and positioning the jack. Realistic jack model with believable textures.
Step 2	Lift the jack by pumping.	Immersion & Expressivity.	Dynamic Interactions	Vibrotactile,	Spatial texture while pumping the jack		Pumping sound coherent with the action
Step 3	Take off the cover from the wheel that protects the bolts.	Immersion	Object manipulation large (two-handed)	Resistive Force Feedback	On grab stiffness to the maximum for both hands. God hand pseudo haptics		Realistic sound, model, and texture.
Step 4	Take the wheel gun. Check if it is in reverse and loosen the bolts.	Realism & Skills Transfer.	Active objects (fast), & Object Manipulation (small)	Vibrotactile, Resistive Force Feedback & Passive Haptics/Skin Indentation.	On grab stiffness to the maximum for manipulating hand. God hand on pseudo haptics. Vibrotactile event on reverse check. Vibration on usage.		Realistic model and texture. Realistic sound for the reverse control. Realistic sound during bolt loosens usage.
Step 5	Wiggle off the wheel and lay it under the car.	Immersion & Transparency	Object manipulation large two-handed)	Resistive Force Feedback	On grab stiffness to the maximum for both hands. God hand pseudo haptics.		Realistic sound when removing the wheel. Realistic model and texture.
Step 6	Take the new wheel and hang it on the pins.	Immersion	Object manipulation large (two-handed)	Resistive Force Feedback	On grab stiffness to the maximum for both hands. God hand pseudo haptics.	*Device-specific SDK* Identify the prefab or preset which applies to the interaction	Same as step 6

Step 7	Take the wheel gun, press the reverse button, and fasten the bolts.	Realism & Skills Transfer	Active objects (fast) & Object Manipulation (small)	Vibrotactile, Resistive Force Feedback & Skin Indentation.	On grab stiffness to the maximum for manipulating hand. God hand on pseudo haptics. Vibrotactile event on reverse control. Vibration on usage.	and include it in the scene. If no prefab or preset is available, it has to be scripted.	Same as step 4
Step 8	Put the cover back into pace.	Immersion	Object manipulation large (two-handed)	Resistive Force Feedback	On grab stiffness to the maximum for both hands. God hand pseudo haptics.		Same as step 3
Step 9	Take the old wheel from underneath the car.	Immersion	Object manipulation large (two-handed)	Resistive Force Feedback	On grab stiffness to the maximum for both hands. God hand pseudo haptics.		Same as step 3
Step 10	Lower the jack, and stuff it away.	Immersion, Expressivity & Transparency	Dynamic Interactions, Object manipulation large.	Vibrotactile & Resistive Force Feedback.	Vibrotactile while interacting with the control, on grab stiffness to the maximum when manipulating		Realistic sounds for lowering the jack
Key deliverables	Project objectives	Haptic goals	Prioritized and consistent haptic interactions	Hardware selection	Game engine & haptic integration tool choice	Interaction framework choice	Realism level vs. investment
Notes	Saving costs compare to real-life training. No extra training is needed after the VR training.	Realism & skills transfer when the wheel gun is involved. Immersion, expressivity, and transparency in any other situation.	Every object should have a Static Interaction (Consistent) Active objects (fast) are prioritized.	Gloves with (resistive) force-feedback and vibrotactile feedback. A tracked dummy prop (pseudo haptics) to secure the skin indentation for the wheel gun.			

11.2. Chapter 4 deliverable: creating your XR scenario

The first step to creating compelling haptics for the XR scenario is describing all the interactions in the scenario. Try to describe the scenario as detailed as possible. If you are creating an enterprise XR training scenario, there are often training manuals at hand. These are ideal documents to start your XR scenario from. In this first phase, it is also key to state the scenario's key business objective(s).

In our example, we will start with a short scenario. The scenario that we take is an XR training for replacing a car wheel with a new type of pneumatic wheel gun (see Figure 24). The main objective of the simulator is to train staff on the new type of wheel gun. Our business objectives are:

1. To save costs compared to real-world training.
2. To fully replace physical training with the new device. No extra training should be necessary after completing the series of virtual training.

To explain our reasoning, we take 3 representative steps from existing real-world training as an example (refer to the XR Scenario column in the training scenario table):

- Step 2: Lift the jack by pumping
- Step 4: Take the wheel gun. Check if it is in reverse and loosen the bolts
- Step 5: Wiggle off the wheel and lay it under the car

11.3. Chapter 5 deliverable: Defining the haptic goals

Write down the haptics goal for each step in the second column. Keep the primary business objectives in mind. This exercise is crucial to design an experience with the right level of fidelity. Well-defined haptic objectives will help to meet the scenario objectives cost-effectively.

When all scenario steps have a haptic objective, they can be summarized. For instance, they can be grouped as in the example scenario. "Realism &

skills transfer when the wheel gun is involved" and "Immersion, expressivity, and transparency in any other situation".

If we take the scenario steps from our example, we define the haptic objectives for step 2 as Immersion & expressivity. Step 2 (lifting the jack by pumping) is not a key interaction in our scenario. We also assume that our trainees are familiar with this procedure. Therefore, we are not aiming for realism here, but we want the trainees to recognize that this step is needed and anchor it to their prior real-world experience. The experience should not break their immersion in the training and should be recognizable for them.

Step 4 is totally different since it represents the core of our virtual training, and our objective is that no further physical training is necessary. Therefore, this step should be as realistic as possible. At the very least, the real skills of handling and configuring this tool should be transferred during this step in the scenario.

Step 5 is again very generic; the sense that it is safe to assume that users have prior experience removing a wheel. Also, this varies depending on the vehicle, but the precise feel of this is not important for the trainer. Therefore, only immersion is chosen. The trainee should believe in the experience. One of the main concerns in this step that could break their immersion would be the constraints of a haptic device. The trainee needs to move a lot during this task. Therefore, transparency is added to the haptic considerations.

11.4. Chapter 6 deliverable: Sketch haptic interactions

When we have our haptic objectives, we need to sketch all the haptic interactions in the scenario. This is a pivotal step to complete before beginning any software or system. This step should provide a sketch of the key mechanics that need to be developed or incorporated from 3rd party interaction frameworks.

In the sample scenario, step 2, the user must lift a jack by pulling on a lever. This is classified as a dynamic interaction. It means that the sensor data for the user's manipulating hand will be used to calculate the haptic feedback. In scenario step 4, the user performs active manipulation (the trigger of the wheel-gun) and object manipulation (catching the bolts). This means that precise finger motion will trigger or dynamically generate haptics. In step 5, the user is performing a two-handed object manipulation. The wheel needs to be grabbed with two hands, and this means that the haptic feedback will need to be triggered when the user's avatar is attached to the wheel object.

11.5. Chapter 7 deliverable: Haptic technology map

The next step is to evaluate which haptics technologies can satisfy the mechanics and fidelity identified in the previous two steps. It is possible that no single haptic device can fulfill these requirements, and then the team needs to decide how to partition the requirements across hardware devices or remove/reduce certain requirements.

In step 2, we need to lift the jack. The main objective at this stage is immersion and expressivity. Triggered vibrotactile feedback should be sufficient when lifting the jack. If the development team knows that the 3D tracked data for the user's hand has high quality, a dynamic effect could also be used.

Step 4 main objective is realism. For this critically important goal, a large palette of haptics sensations is useful. Vibrotactile feedback is needed to simulate the power-on state of the wheel-gun. Resistive force feedback is necessary to simulate the trigger button of the drill, and passive haptics/skin indentation is needed to identify the shape of the object.

In step 5, resistive force-feedback should be sufficient. We don't want to break the user's immersion by letting them grab a wheel in thin air. In this case, a rough outline of the wheel's size will be more than sufficient. Vibrotactile feedback can also achieve this since realism is not a goal in step 5.

In the example scenario, a wearable resistive-based force feedback glove with a dummy pseudo-haptic wheel-gun was chosen. This configuration provides excellent coverage over the haptic goals and enables all the interaction mechanics with high transparency.

Another pricier option could be a wearable resistive force feedback device equipped with pneumatic contact spatial actuators. However, this would be complex to deploy and would have poor device transparency compared to the previous solution.

11.6. Chapter 8 deliverable: Haptic design choice

Now that the decision is made on the hardware, the team can design the desired interaction primitives. These interaction primitives are the primary means for guiding the integration of interactions and haptic effects into their scenario. Where in previous steps, the interactions and haptics were separated from each other. It is necessary to examine how the interactions drive the haptic effects.

In step 2, the team decides to create a spatial texture for pumping the jack. This is accomplished with a continuous dynamic vibrotactile signal where the frequency and amplitude depend on the angle of the jack handle. The chosen haptic hardware allows for HD haptics and, therefore, can simultaneously vary both speed and amplitude of the waveform. This is implemented using a signal generator updated at the collision/sensor update rate.

In step 4, the team leverages the pseudo-haptic prop for most of the haptic interactions. It is, however, crucial that the visual virtual hand is rendered correctly with the usage of a god hand illusion. This means that the visual hands never intersect the virtual object. This visual illusion, combined with the pseudo-haptic object's shape, creates an extremely convincing sense of realism for users. Furthermore, when the trigger is pressed, the team generates a dynamic resistive kinesthetic signal that increases, corresponding to the trigger displacement. At last, a dynamic vibrotactile feedback signal is necessary where the amplitude is dynamic to the penetration dept of the trigger to simulate the feeling of a real tire gun.

In step 5, only a dynamic passive kinesthetic signal is needed, which can always be at 100% strength. The primary consideration here is to ensure that the resistance of the two hands is consistent with how the user has grabbed the virtual tire. To render the object nicely within the hands, a god hand illusion is also preferred here.

Once the interaction haptics have been designed, the team can incorporate this information into the selection process for the overall simulation engine. As discussed in section 8.1, there are many excellent options. Considering the choice of hardware and the available integration leads the team to select Unity since there are excellent interaction development assets and support for resistive glove-type devices available for this platform.

11.7. Chapter 9 deliverable: Prototype interactions

It is time to start prototyping the interactions and haptic effects. From the deliverable in steps 11.4 and 11.5, the team now has a clear idea of what to prototype. Step 11.6 provides the guidance to implement the haptic design. Prototyping requires a choice of haptic effect creation tool(s). This can be a device-specific or generic tool. The choice of haptics and interaction tool will be mainly based on two factors. The first factor is the amount of custom development work. Many tools come with pre-build interactions and tools to program your haptic effects. The more these pre-build interactions match the chosen haptic interactions in the scenario, the easier the integration becomes. When there is less overlap with the chosen tool, the team should expect to have to create custom code to realize the haptic effects. The second factor is the choice of tool-specific SDK or generic tools.

In the tire-gun scenario, the chosen hardware is well supported by a popular haptic authoring tool available for Unity (Interhaptics), and this tool can be used to prototype and implement all the designed interactions.

A good practice for haptic effect creation is to make sure the prototype interactions work as expected first with basic interactable objects (just virtual boxes that match the size and shape of the final objects) without detailing. Before detailing the visuals and audio, the game state interactions and haptic effects should be finished first.

Doing this makes it easy to iterate on the interaction and haptics. If all interactions are working and tested, then detailed visuals can be added at any point. Otherwise, the visual detail may interfere with or block interaction-specific optimizations.

11.8. Chapter 10 deliverable: Integrated multi-sensory scenario

The last part is to finalize the scenario with multisensory information. The team considers the first deliverables as the first step in an iterative process that should be repeatedly evaluated with stakeholders and representative users before finalizing. Tuning the multi-model integration typically needs one or two iterations with the haptic design and development tasks before it works well. The multi-sensory implementation should be done via the game engine where the XR environment is created in. This includes but is not limited to (spatial) audio, visual textures, visual cues (such as arrows and pointers), detailed shaped 3d objects, and multisensory dependencies. Three considerations can optimize the haptic experience:

The first consideration is the detail of the mesh used for collision; it should match the physical shape as closely as possible. This is necessary because most haptics feedback will be utilize a mesh collider collision engine. It is also possible to simplify colliders with a set of bounding spheres and box colliders to improve latency, but the mismatch between the visual and collision geometry may confuse users. For the wheel gun, the hand must look very realistic as it manipulates the tool. For this reason, the resolution, textures, and sounds of the gun itself should be as high fidelity as possible.

A second consideration that could make a better haptic experience is to dynamically modify the visual deformation of the material. If there is a soft object with dynamic **kinesthetic** feedback, a deforming visual of that object corresponding to your dynamic effect helps a lot in the user's immersion. For example, if the tire being replaced has low air pressure, the visual rendering should deform, and the dynamic stiffness haptics should provide complementary tactile feedback.

The last consideration is to add task completion haptics (non-diegetic), perhaps along with a visual message and/or audible tone. The developer can include gentle, rewarding haptic signals when successfully executing a task or game state. For example, once the user can remove all the wheel nuts, they might get a gentle haptic pulse with an audible chime, letting them know they have succeeded.

11.9. Testing and Iteration

Integrating haptic feedback into an XR environment is partly an art and a science. This book provides a systematic methodology for executing and delivering high-quality, high-value haptic sensations. Just following the guidance in this book is insufficient, however. It is critical to test and iterative the various mechanics and the entire scenario with users and stakeholders as often as possible. Haptics is a sensation that many users have a hard time articulating and specifying with words. It is typically necessary to have the ultimate customer have a visceral experience for them to be able to provide guidance and feedback.

12. Upgrading and Maintenance

Much of this guide has focused on evaluating and implementing haptics as part of the overall simulation development. In this chapter, the authors provide some thoughts on upgrading an existing virtual reality system that does not already include haptics. There is a wide range of complexity and cost associated with this retrofit, and special planning may be necessary. Depending on the size/scope of the project, it may be best to engage haptic XR specialists or consultants before doing this.

A legacy virtual reality system based on a custom or proprietary 3D engine might not be capable of the frame rates or collision calculations required for immersive, dynamic haptic effects. In general, static effects should be used for legacy systems with low or unreliable simulation frame rates. If dynamic haptic effects are needed to realize the desired haptic value, it may be necessary to rearchitect the simulation significantly. It may also be possible to move the haptic interaction processing loop to another processing thread or another computer and use the low frame rate legacy simulation to provide low-frequency updates. Ensuring consistency between visual and haptic feedback can be challenging in this architecture.

In case the Virtual Reality scenario was built in a recent version of Unity 3D (2018 +), there is a faster and more accessible approach to upgrade the virtual reality scenario because Unity already incorporates a sufficiently sophisticated interaction framework.

Implementing a haptic output like a haptic vest is straightforward through the use of device-specific SDKs, audio pipelines, or haptic-specific rendering software like Interhaptics. The haptic pipeline is separated from the content and does not change the structure of the scenario.

The process becomes a little more tedious with substituting the legacy interaction systems such as VRTK and others with a haptic interaction system like Interhaptics or device-specific SDKs like Senseglove SDK. In this case, the interaction logic and scripts should be substituted with hand interaction and haptics enabled scripts. The process is not complex but requires rebuilding the interactions from scratch to be compatible with the integrated haptic frameworks.

In summary, updating an existing simulation is achievable if the system already has a sophisticated interaction engine and the haptic feedback consists primarily of static effects. Legacy systems that do not have high frame rate collision detection should be approached cautiously and with domain experts to avoid unacceptable results.

Part II: Use Cases and Applications

Chapters 13, 14, and 15 focus on the real-world implementation of the concepts shared in the previous chapters. The focus, in this case, is on Enterprise VR applications due to the historically non-consumer-friendly cost of haptics systems.

Chapter 13 is focused on application areas. The product or program manager will find helpful information in these applications to plan the value of haptics for the project or product stakeholders.

Chapter 14 shares a list of use cases related to the application areas shared in the previous chapter.

Chapter 15 proposes implementation examples for the application areas with a critical discussion of the suitable haptics technologies to meet the use case and haptics objectives for the product/project.

13. Enterprise XR Application Areas

Haptics have been utilized in XR applications for several decades. It is worthwhile to review the common application areas where the inclusion of haptics has proven value and extract learnings that can guide product planners and system architects. This chapter summarizes the common application areas for haptics in XR and provides a template for thinking about the return on investment of haptics. The next section presents a series of case studies that connect these application areas to real-world simulation products.

Each application area will be described using a template that explains the scope, cost, haptic and non-haptic considerations, how to think about success, as well as some application-specific suggestions to ensure success.

13.1. Application Area: Virtual Prototyping

13.1.1. Overview

Virtual prototyping involves simulating object behavior under real-world operating conditions; therefore, the sensorial experience is as important as the visual one. Virtual prototyping is commonly used in enterprise scenarios, particularly when the cost of changes to a design increases exponentially as the product passes each design gate. Examples include interior design for cockpits of aircraft, passenger vehicles, and other infrastructure where human-machine interfaces will be complex and are challenging to evaluate with physical prototypes. XR systems provide early-stage cost and complexity advantages and enable rapid iteration during early design stages. When the eventual product has a meaningful tactile dimension, the addition of haptics can increase realism and enable better judgement of the design alternatives.

13.1.2. Investment Range

Consumer ($) to Enterprise simulative training ($$$$)

13.1.3. Non-Haptic Considerations

Application maturity: A successful virtual prototyping application requires the user to be familiar with virtual reality technology and have successfully delivered proof of concepts validating the ROI.

Developer/Integrator capabilities: The developer and integrator capabilities are essential factors to consider while engaging with VR prototyping content creation. VR prototyping applications are well served with refined accessibility features.

CAD to VR pipeline: One of the key aspects of virtual prototyping is the implementation pipeline for CAD data to enable the review process.

Multi versus single-player: Shared experiences help users speed up the virtual prototyping process through real-time communication.

13.1.4. Customer Goals

- Reduce prototyping costs through savings on physical prototypes and mock-ups. Reduce cost compared to traditional digital prototyping tools like caves and powerwalls.
- Speed up development cycles.
- Standardized solution: Have one generic solution for all digital prototyping efforts.
- Dispersed Locations: A VR solution is portable and can be positioned anywhere, even in a home environment. VR offers multiplayer solutions where several engineers can be present during the assessment of the digital prototype.

13.1.5. Role of Haptics

Realism. A similar behavior between the virtual and real environment facilitates an informed understanding of human behavior. The improved perceived realism makes it easier for the user to make decisions on the UX, manufacturability, or ergonomics of a virtual prototype.

Immersion. Enabling a more realistic and engaging testing environment. Haptics can enhance the users' immersion during the execution of a test.

User experience. Haptics or hand tracking enables an interaction that doesn't require a learning curve, rather than controllers where every interaction is programmed differently. It enables the opportunity to work with untrained users. This can be critical in testing virtual prototypes on user behavior.

13.1.6. Haptic Considerations for Success

Wearability. Virtual prototyping needs to meet operational targets to justify its ROI. The haptics device used for these applications should be easy to wear comfortable, and it should not impede users' natural movements.

Embodiment. The embodiment of the haptic technology should be fit for the right purpose. Is the objective to perform simple ergonomic analysis or to provide designers with a tool to evaluate a prototype from a remote space? Or is the objective to digitalize the objective of a clay model in the automotive design process? Both situations require a different embodiment. The solution's transparency should be based upon the desired fidelity and the associated ROI of the solution.

Scalability (hours of use, number of deployments, etc.). One of the key aspects to consider when thinking about haptics in VR is where the scalability comes from. Do you need repeatability of the same content on different premises to target hundreds if not thousands of users? Or are you looking to create a one-of-a-kind setup for critical operation training for specific applications? Both scenarios have different needs and requirements that different haptic devices can address.

Integration of non-haptic peripherals and mock-ups. Mixed reality mock-ups are an excellent way to iterate in prototyping rapidly. Consider whether your haptic solution can interact with the physical part of the mock-up and whether it can simulate the right fidelity of haptics.

Content enablement. The trade-off that should be considered is the ease of implementation against the quality of haptics. Standard haptic interaction might be compatible with direct CAD to VR pipeline solutions; more complex haptic interactions require tailored integrations.

13.1.7. Use Cases

In section 15.5, a Virtual Prototype use case is reported. Fraunhofer IEM has created an Augmented Reality application with a haptics exoskeleton in a joint research project with Hella Headlights. With a combination of a paper mock-up of an assembly cell, an AR headset, and the haptics device, Hella optimized their design for manufacturing processes in the early product development stage.

13.2. Application Area: Training

13.2.1. Overview

Training is one of XR's foundational application areas and one of the earliest to utilize haptics in a demonstrably value-added manner. XR environments are extremely adept at enabling training and simulation of real-world scenarios, particularly ones that are risky or hard to duplicate. This enables systematic skills training with complete control over the distribution of trainable events. Typical use cases include assembly/disassembly training, surgical procedure training, firefighter training, and military training.

13.2.2. Investment Range

Consumer ($) to Enterprise simulative training ($$$$)

13.2.3. Non-Haptic Considerations

Application maturity. A successful virtual training application requires the user company to be familiar with virtual reality technology and to have successfully delivered proof of concepts validating the ROI.

Developer/Integrator capabilities: The developer and integrator capabilities are important factors to consider while engaging with VR training content creation. VR training applications are well served with great user experiences, refined accessibility, and usability features.

13.2.4. Customer Goal(s)

- Reduce operational costs of delivering practical training:
    - Reduce real equipment usage for training purposes
    - Reduce maintenance costs for training centers
    - Reduce travel cost for trainee
- Ensure repeatability thanks to the digital format of the learning support
- Ensure learning consistency thanks to the digital support
- Increase user retention through increased immersion and gamification techniques

13.2.5. Role of Haptics

Skills transfer. Well-designed applications, including haptic feedback in VR training, can generate positive learning reinforcement, enhancing training effectiveness. This results in a lower rate of error during the training process.

Realism. A similar behavior between the virtual and real environment facilitates skill transfer between the virtual and real case. The absence of natural interaction and realistic haptics can generate bad practices or negative learning that must be unlearned in real-life skill implementation.

Immersion. Haptics can enhance user immersion during the execution of the training content. User immersion increases the embodiment and believability of the training scenario, increasing its effectiveness.

User experience. Hand tracking and natural interactions coupled with well-designed haptic feedback can meet or exceed user expectations of interactive content. Interactions can be more precise and realistic, thus reinforcing the sense of presence. This reinforces the "learn by doing" value of VR training. The hand-tracked interaction is transparent and natural, whereas a metaphor delivered by the system mediates the controller interaction.

Ergonomics/usability. VR training is accessible to everyone through the use of hands. The use of haptics reduces the friction of adopting VR training content.

13.2.6. Haptic Considerations for Success

Wearability. Virtual reality training needs to meet operational targets to justify its ROI. The haptics device used for these applications should be easy to wear, comfortable, and should not impede users' natural movements.

Embodiment. The ideal haptic device should be transparent while not interacting and perfectly reproduces reality while interacting with VR. System developers should optimize transparency, given their budget constraints.

Scalability *(hours of use, number of deployments, etc.).* One of the key aspects to consider when thinking about haptics in VR is where the scalability comes from. Do you need repeatability of the same content on different premises to target hundreds if not thousands of users? Or are you looking to create a one-of-a-kind setup for critical operation training for specific applications? Both scenarios have different needs and requirements that different haptic devices can address.

Skills training: hard or soft? Are you planning to teach how to use a simple HMI system, ease the use of a soft skill training scenario, or train how to mount an aircraft precisely? These use cases require different levels of haptics, starting from optical hand tracking towards a high-fidelity hand haptics device.

Integration of non-haptic peripherals. Increase the training fidelity by passive haptics with real "dummy objects" with haptic feedback from the device.

Content enablement. Implementing haptics in existing content requires the specification of the haptics experience to be created. This can be delivered by purchasing existing haptics assets to be combined into the scenario, using haptics design software to create the necessary content, or with the support of a service provider delivering specialized haptics development.

13.2.7. Use Cases

Several virtual training use cases are reported in sections 14.1, 14.2, and 14.5. Ranging from electrical maintenance training operations to automotive assembly training, automotive painting training, satellite assembly training, and realistic first aid training. The role of haptics is focused on accessibility and skill transfer. The outcome is usually positive, and these use cases are in use to this day.

13.3. Application Area: Marketing/Sales

13.3.1. Overview

Virtual reality technologies have the capability to showcase products and experiences that are impossible or extremely costly to build. It also enables customers to have a virtual design experience for products that are customizable before they are manufactured (e.g., yachts, aircraft, etc.). It has applications for trade show experiences, in-store experiences for retail, experiential marketing, and to raise user awareness for specific topics.

13.3.2. Investment Range

($) Consumer to ($$$) high-quality production

13.3.3. Non-Haptic Considerations

Application maturity. A successful VR marketing application requires the marketer to be familiar with virtual reality technology and to have successfully delivered proof of concepts validating the ROI.

Developer/Integrator capabilities. The developer and integrator capabilities are important factors to consider while engaging with VR marketing content creation. VR marketing applications are well served with great visual content and refined experiences.

Asset Compatibility: 3D source material should respect the compatibility necessities for real-time haptics interactions. This is especially relevant when re-using pre-existing assets.

13.3.4. Customer Goals

Engage the user. A well-designed VR marketing experience generates a deep sense of presence, increasing user retention and engagement during the experience.

Increase customer conversion rate. User immersion within virtual reality marketing applications can increase customer conversion. With VR applications, marketers can enhance user immersion and content interactivity. Marketing applications result in a greater lifelike experience and testing capabilities for customers. This is especially relevant for applications in real estate or complex products like vehicles, tool machinery, or boats.

Speed up the sales process. VR applications can help the sales of complex or spatially large products. The more significant benefits happen when a product requires user testing or live presence to evaluate human factors.

Reduce marketing costs. Virtual reality marketing applications can be realized once, and deployed on multiple sites, easily transported, shipped, and deployed in front of the customers. An entire catalogue of complex products can fit into a portable headset.

13.3.5. Role of Haptics

Realism/Differentiation of active haptic elements. Haptics can differentiate and bring closer to reality product aspects and characteristics necessary for the marketer. Demonstrating how an HMI system of a virtual product responds to the touch or how a lever activating a certain command can be delivered with the right haptic technology. The user will gain a visceral understanding of how the physical product would behave.

Immersion. Well-designed haptics enhances the user immersion during the marketing activities. The user will feel more present and in control of the situation, resulting in greater retention and engagement during the experience. In the end, greater customer engagement means a greater chance of a conversion.

Accessibility. Haptics guarantees that the virtual reality marketing content is accessible like the real product. Users simply use their hands, feel the feedback of the virtual environment, and act much like in reality.

13.3.6. Haptic Considerations for Success

Wearability. Virtual reality marketing needs to meet operational targets to justify its ROI. The haptics device used for these applications should be easy to wear, comfortable, and should not impede users' natural movements.

Embodiment.
- What is the continuum between passive haptic realism and simulated active haptic from a cost and fidelity perspective?
- Marketers must consider the purpose of their content and the target user. They must balance the passive haptics effect, generating a believable albeit incomplete sensorial experience, with the stronger embodiment and more profound experience delivered by simulated active haptics. There are existing haptics

solutions balancing passive haptics capabilities vs. fully simulated haptics. The material, integration, and experience design budget usually grow with higher experience fidelity. Marketers should define the success criteria for the experience and derive the best haptics solution.

Scalability. Marketers need to identify what determines the scalability of their content. Will the solution be used by VIPs in high-end retail stores or realized in hundreds of units and shipped to customers? In both cases, the right haptics can fit with the required scalability.

User experience (WOW effect). Great haptics elicit the WOW effect among potential users. This is a powerful tool for marketers to substantiate sensorial marketing applications and make the experience memorable.

Content enablement. Implementing haptics in existing content requires the specification of the haptics experience to be created. This can be delivered by purchasing existing haptics assets to be combined into the scenario, using haptics design software to create the necessary content, or with the support of a service provider delivering specialized haptics development.

13.3.7. Use Cases

In section 14.6, a representative use case for the marketing of nerve damage therapies. This use case achieved the customer's goal of increasing engagement and enabling the marketing of a very novel experience. The haptic component of this use case needed to be usable in a trade-show environment, be scalable, and create a WOW effect for the target customers.

13.4. Application Area: Tele-existence/Tele-robotics

13.4.1. Overview

Tele-existence is the ability to reproduce human capabilities in another physical avatar. Tele-existence is advantageous in dangerous situations/environments, sterile environments, and geographically distant locations. The advantages of tele-existence are the ability for real-time expertise, interactions, and oversight. This can vary from CBRNE (Chemical, Biological, Radiological, Nuclear, and high yield Explosives) to maintenance and medical use cases. Tele-existence systems may consist of at least two of three technological components:

1. The ability to reproduce visual and audio information; this is usually referred to as tele-presence or tele-existence.
2. The ability to create mobility referred to as locomotion.
3. The ability to convey tactile transfer and physical interaction between the robot and the human. The human controls the robot's actions, and the robot accordingly responds to the human's actions and conveys tactile sensation to the human.

13.4.2. Investment Range

Consumer ($) to clinical facility ($$$$)

13.4.3. Customers Goals

Reduce risks. When dealing with scenarios like CBRNE or dangerous maintenance tasks, exposing a robot to such danger is ethically more responsible than exposing a human to such a threat.

Skill teleportation. Using a robot as a personal avatar, a human may impart their expertise through the robot to a remote location or a location that is difficult to access through the human form without mechanical assistance.

Increase efficiency. Telerobotics allows humans to avoid dangerous, distant, and dirty environments. It enables the human expertise to be transmitted through the telerobot, potentially eliminating the need for travel while still providing a sense of first-person interaction and expertise through the robot.

13.4.4. Role of Haptics

Increased realism. Haptic feedback from the telerobot to the human provides a greater sense of realism embodied in the robot, a greater sense of dexterity and proprioceptive presence to the human through the robot, and the familiarity of natural movement and task conduct through haptic feedback, which acts as confirming sensations for task conduct and completion.

Increased performance. Using telerobotics/tele-presence, one person can be present in many geographically dispersed environments through the deployed robotic avatar. This person's expertise and know-how are conveyed through the telerobot, enabling the robot to complete the task with the expert's ability.

Provide safety and control. Sparing first-person exposure to dangerous situations or environments, haptic feedback through robotic sensors provides real-time feedback and a natural human interface for task completion with high levels of confidence and familiarity with the conduct of the telerobotic avatar.

13.4.5. Haptic Considerations for Success

Realism: Tele existence is well served by realistic haptics sensation to optimize the remote control of the robot. Haptics realism should be actively pursued, and it is a critical item for success.

Safety: Remote controlling robots implies potential user safety risks. The haptics implementation should put the user safety first and avoid any haptic feedback above the user tolerance.

Integration: Integrating an external robotic system is challenging and requires a specialized approach.

13.4.6. Use Cases

Haptic teleoperation is a commercial feature of surgical robots, drone operators, and other high-risk use cases. Telerobotic systems have special performance requirements often dictated by the specific configuration and connectivity of the remote robot manipulanda and the user.

13.5. Application Area: Assistive

13.5.1. Overview

Assistive use cases relate to health and wellness, rehabilitation, and other typically therapeutic or clinical uses to improve patient outcomes. This area is relatively nascent in terms of haptics but does provide some novel considerations relative to the use and value of haptic feedback. Virtual reality technologies are used within rehabilitation processes for impaired patients. Virtual reality targets specific movements or subjects the patient to particular stimuli to accelerate recovery.

13.5.2. Investment Range

Consumer ($) to clinical facility ($$$$)

13.5.3. Customer Goals

The goals of these use cases are typically to influence or generate a measurable improvement in patient responses to standard measures. Measures can include physiological improvements such as reduced heart rate, anxiety, etc., task-related improvements such as task completion times, error rates, etc., or other miscellaneous measures such as therapy duration.

Reducing costs. Well-designed VR experience can enable unsupervised training. The virtual training task should be self-explanatory. In that way, multiple patients can train in groups or at home with just the guidance of one physical therapist.

Increased engagement. VR can help improve the engagement of patients in their therapy. It can be adaptive to the challenge of the patient, providing a challenging but achievable goal. Also, the knowledge of being monitored increases therapy compliance.

Improved insights. VR therapy can increase the level of insight on progress and performances within the physical rehabilitation process.

13.5.4. Role of Haptics

The specific role of haptics varies materially by use case, but generally speaking, it can have the following values:

Generation of force/resistance for muscular training/rehabilitation use cases. Increasing the resistance in a training bicycle is a type of haptics; moreover, this can be associated with a virtual reality experience.

Generation of error feedback for body pose, task progression, or other training/rehabilitation-specific measures. Posture correcting yoga pants (e.g., wearable.x) in conjunction with a VR yoga class is an example use. Note that the haptic feedback can be automatic or triggered by a coach/trainer in this scenario.

Generation of increased immersion. Similar to other broad use cases, haptics is known to increase the immersive quality of a virtual reality experience. This can be valuable for training or entertainment-driven assistive use cases. For example, providing content-synchronized feedback for non-ambulatory users (e.g., in a hospital bed) can increase the quality and duration of the engagement.

13.5.5. Haptic Considerations for Success

Generally speaking, assistive haptic feedback needs to consider these key points:

Safety. Devices that generate forces on users need to be engineered so that they cannot cause injury due to inadvertent use. For rehabilitation use cases, this consideration is critical as users may have non-normal ranges of motion, sensitivity, etc. If possible, users and practitioners should be empowered to reduce/eliminate the haptic feedback in real-time to allow individual variation.

Consistency. In order to use haptics to generate precise, measurable outcomes, the haptic feedback must have a maximum amount of consistency both from use-to-use and from installation-to-installation. For body-mounted feedback, careful consideration should be paid to fit consistency, including having multiple sizes or specific criteria for appropriate body size considerations.

Perceptibility. In addition to consistency, haptic feedback needs to be noticeable and provide the appropriate dynamic range of sensation. This is particularly relevant to error-driven haptic feedback. If the specific use case involves vigorous physical activity, haptic sensations typically need to be stronger but may also need to be carefully designed to disambiguate them from other scenario-related stimuli. Overly complex haptic signaling can increase cognitive load and confusion.

Expressivity. Haptic devices need to provide a sufficiently perceptible range of experience so that users can correctly interpret the stimuli. Basic body-mounted error feedback may only need 1-bit of expressivity (on/off), whereas an error measure that indicates the quantity of the error may require many more bits.

Wearability. Virtual reality rehabilitation needs to match therapy or reimbursement budgets. The haptics device used for these applications should be easy to wear, comfortable, and should not impede users' natural movements.

Embodiment. The ideal haptic device should be transparent while not interacting and perfectly reproduces reality while interacting with VR. The ease of use in putting on and removing is key for rehabilitation. System developers should optimize transparency given their budget constraints.

Scalability *(hours of use, number of deployments, etc.)*. One of the key aspects to consider when thinking about haptics in VR is where the scalability comes from. Do you need repeatability of the same content on different premises (for instance, clinical therapy)? Or would you like to target hundreds if not thousands of users (home-based rehabilitation)? Both scenarios have different needs and requirements that different haptics devices can address.

Reliability. A haptic system should be reliable. It should always work, especially in home-based environments where no specialist is available to debug a potential problem.

13.5.6. Non-Haptic Considerations

Security. Assistive virtual reality use cases deal with patient data, which is a regulated area and requires specific management.

Cognition. Virtual reality is a technology that might not fit the cognitive state of some patients.

13.5.7. Use cases

Though still in the early stages, haptic feedback has tremendous potential value for rehabilitation and physical therapy. In an XR context, assistive haptic technologies should enable faster and more successful recovery and enable higher functioning of users.

14. Use Cases

This chapter summarizes specific, real-world use cases that have been developed by the authors or contributors to this publication. These use cases are offered to ground the recommended practices and inspire readers with the possibilities and achieved value of haptics in XR.

14.1. Electrical Maintenance Training

14.1.1. Overview

A well-known electrical equipment provider commercializes low and medium voltage needing regular maintenance. The customer's workforce needs to perform scheduled training to learn and refresh the maintenance operations procedures. This training is performed at training centers around the globe. The sessions are costly, involving the travel of the workforce to the training center, a few days stay to conduct training activities on dummy machines under the trainer's supervision. The electrical equipment manufacturer developed a haptics training solution to digitize the maintenance and security training for the workforce to bring the training sessions to the customer. The solution also commercializes a VR training system, including haptics and VR equipment, to allow their customer to keep the training scenarios and experience as documentation.

Figure 25: Electrical equipment maintenance and training system allow trainers to train on novel scenarios.

14.1.2. Customer Goals

Risk reduction, skill transfer, reduced costs, increased margins.

14.1.3. Project Budget

>$100,000 USD

14.1.4. Haptic Technologies Used

The system was developed in Unity and incorporated 4 VR Touch (Go Touch VR) gloves. These gloves provide skin indentation feedback to enable tactile sensations related to grasping and manipulation.

14.1.5. Role of Haptics

Haptics was added to the simulation to increase usability and skills transfer. It was initially challenging to communicate this value to the customer. Once the customer understood that haptics does not reproduce reality with perfect fidelity, it became easier to have a pragmatic discussion about the correct use of haptic feedback in the simulation.

14.1.6. Outcome

The project continues to be used, after two years, by the customer. The enhanced usability created with haptic feedback is a key reason for the lasting success of this use case.

14.2. Automotive Painting Training

14.2.1. Overview

VR-based training application for automotive painting. A range of vibrations were introduced to guide the user through the painting processes, errors, and the set of guidelines. If the applied pressure is high, the user gets particular feedback. If the applied pressure is low, the user receives the appropriate feedback. An entire set of feedback is utilized for different sections of the training guidelines.

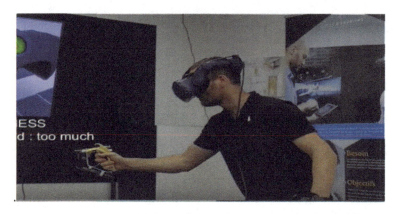

Figure 26: The custom haptics interface allows a realistic experience while training for automotive painting tasks.

14.2.2. Customer Goals

Risk reduction, skill transfer, product/experience insights.

14.2.3. Project Budget

$10,000-$100,000

14.2.4. Haptic Technologies Used

The type of haptic feedback used for the project was based on vibrotactile perception. The actuators were Actronika's patented voice-coil motors (Hapcoil One) with a resonant frequency of 65 Hz. These VCAs have a form factor of 11.5x12x37.7 cubic mm and have the potential to operate with promising results in the range of 10 Hz to 1 kHz. The acceleration ranges between 8 g-pp to 11.4 g-pp depending on the type of signal sent. The VCAs were driven by our embedded solutions, essentially a software-based control (Unitouch Embedded). The vibrations used were part of our library of UI effects.

14.2.5. Role of Haptics

Immersion, usability, user experience.

14.2.6. Outcome

The product is in use by the customer who is ordering parts every quarter and is listed as one of the most satisfied customers.

14.3. Satellite Assembly Training

14.3.1. Overview

SenseGlove Nova has been implemented in the military VR training for assembling a satellite receiver to avoid damaging expensive training equipment. Using haptics gloves in a virtual environment, the trainees can install the virtual parts of the satellite receiver in a similar way as they would install the real equipment.

Figure 27: SenseGlove and VR headset for satellite assembly training.

14.3.2. Customer Goals

Skill transfer, reduced development costs.

14.3.3. Project Budget

$10,000-$100,000

14.3.4. Haptic Technologies Used

The SenseGlove Nova force and haptic feedback gloves. A multiplayer game based on the Unity engine and in integration solution of VREE, a Dutch VR development agency.

14.3.5. Role of Haptics

Realism, immersion, user experience.

14.3.6. Outcome

The study is not finished yet after 4 months of activity, but the first results are positive. There need to be some tweaks, but it has a high chance of becoming the standard way of interaction for this training of the Dutch defense.

14.4. Discover History with Haptics

14.4.1. Overview

Transforming education and history research with haptics. The study explores how haptic technologies combined with VR can better engage viewers in the museum and heritage sectors.

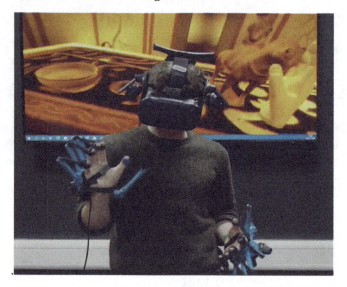

Figure 28: Discover History with the haptics use case. The user can touch and manipulate virtual art objects in virtual reality.

14.4.2. Customer Goals

Product/experience insights.

14.4.3. Project Budget

$1000-$10,000

14.4.4. Haptic Technologies Used

SenseGlove DK1 and virtual reality headset.

14.4.5. Role of Haptics

Realism, immersion, user experience.

14.4.6. Outcome

"The haptics spark the potential to revive how we interact, communicate, and preserve ceramic artefacts long term. According to Emma Fallows, a researcher at Staffordshire University, this innovative way of interaction reinvents the way we communicate history, aids interpretation, and increases visitor engagement". Early tests have shown that VR with haptics may become a popular option for preserving ancient artefacts via a digitized archive. The project is a 4-year PhD study and is still active.

14.5. Electrical Assembly Training

14.5.1. Overview

Volkswagen and SenseGlove have created a virtual reality assembly training of the electric components within the sliding door of the T6 van. In this scenario, participants can train the complete procedure of all actions in this assembly process.

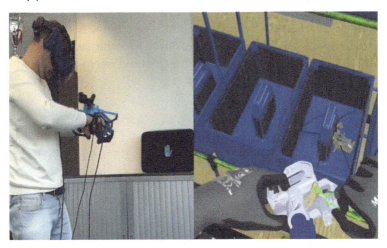

Figure 29: Representation of the assembly use case for Volkswagen realised with the Senseglove.

14.5.2. Customer Goals

Skill transfer, reduced development costs.

14.5.3. Project Budget

$10,000-$100,000

14.5.4. Haptic Technologies Used

SenseGlove DK1. A CAD to Unity pipeline. And the Unity game engine to create the content.

14.5.5. Role of Haptics

Skills transfer, realism, immersion, usability.

14.5.6. Outcome

The project satisfied the objective as a research project. For the desired implementation, a new follow-up project is needed with improved usability. This specific project lasted for 4 months and is finished. A follow-up project is scheduled for 2021.

14.6. Nerve damage experience

14.6.1. Overview

SenseGlove has created a VR environment where nerve damage symptoms can be experienced. This simulation was used in a worldwide campaign for Procter & Gamble Health to raise awareness about nerve damage and build empathy for those suffering from it.

Figure 30: Nerve damage experience

14.6.2. Customer Goals

Product/experience insights.

14.6.3. Project Budget

$10,000-$100,000

14.6.4. Haptic Technologies Used

SenseGlove DK1, Unity plugin by SenseGlove.

14.6.5. Role of Haptics

Immersion, user experience, create empathy.

14.6.6. Outcome

For P&G Health, it is hard to communicate a marketing message of a disease that people and general practitioners commonly overlook. Therefore, they wanted to create a campaign to increase awareness of nerve damage. SenseGlove was the ideal tool for P&G Health. At conferences, they can let GPs experience nerve damage symptoms themselves, so GPs can learn to diagnose these symptoms earlier. This innovative way of marketing also creates significantly more traffic to their stand at conferences and roadshows. The project lasted for a year. It kicked off with a movie in 2019, and the showcased travelled around the world in countries like Brazil, the Philippines, Malaysia, Switzerland, and Portugal. Via the spreading of the video, P&G Health can increase the awareness in public on the symptoms of nerve damage and build their brand effectively and innovatively.

14.7. Augmented Reality Haptics Showcase

14.7.1. Overview

In collaboration with Meta (Headset manufacturer), Ultraleap worked on a vision for the future of design. With the participation of Dell and Nike, they first presented this vision through a short clip, showing a rethink of the designer workspace. Specifically, using today's technology, hand tracking, and mid-air haptic, they can blur the digital and physical worlds and deliver meaningful applications and tools to the workspace.

In this second stage of this collaboration, Ultraleap was joined by ZeroLight and developed the first proof of concept using automotive as the use case. In this demonstrator, Ultraleap replaced all conventional elements of a workspace (screen, keyboard, computer mouse) with an AR headset, a mid-air haptic display, and a hand-tracking sensor so that users could interact and manipulate visual holograms while receiving haptic feedback through the various interactions. A total of 6 scenes composed the demonstrator. In the first scene, users activated the interface by pressing and holding down a mid-air haptic button in the center of the interaction space. In the second scene, users were invited to pick the color of a car from a rotary dial that they could spin. The resulting car visual was presented as a floating hologram in the center of the interaction space. In the third scene, a close-up view of the car engine is shown to the user. Users could hear (audio) and feel (haptics) the engine roar upon touching the motor. In the fourth scene, the car was presented as a floating hologram at the center of the interaction space. When users reach and touch the car's roof, its visuals will change into an exploded view, showing all the components it was made from. This exploded view represented the fifth scene of the demonstrator. If users withdrew their hand, the interaction would revert to the fourth scene, but if users kept pushing their hand down, the elements from the exploded view would be sent around the users at a bigger scale, representing the sixth and final scene. In this last scene, users could pan the different elements around them using their hands.

Figure 31: Example image for AR haptics marketing experience.

14.7.2. Customer Goals

Product/experience insights.

14.7.3. Project Budget

$10,000-$100,000

14.7.4. Haptic Technologies Used

For this project, Ultraleap used ultrasound mid-air haptic technology, which Ultraleap is the only company worldwide to commercialize. This technology uses an array of ultrasound speakers (hence inaudible) to focus acoustic waves at a given point in space and time. Pressure is sufficient to slightly push the skin and induce tactile feedback at this specific point. Controlling this point pressure amplitude and position over time, the display can convey a whole range of haptic effects. The haptic effect can be projected directly onto the user's palm with hand-tracking technology. Here, Ultraleap used a leap motion controller as the hand-tracking solution, which the company also commercializes. Both devices' APIs are compatible with various tools and languages. Specifically for this project, Ultraleap used Unity. For the AR mounted display, the Meta2 glasses were used, produced by Meta (a company that has shut down since), which was also compatible with Unity.

14.7.5. Role of Haptics

Skills transfer, realism, immersion, usability.

14.7.6. Outcome

The primary objective for the project was a working demonstrator, which was completed to the satisfaction of the customer. Further goals, such as integration into a product of production quality, were not pursued. The project lasted six months.

14.8. Medical Care Training

14.8.1. Overview

In 2020, the United States Defense Health Agency awarded a contract to HaptX and others to develop a haptics-based training system to improve the skills of U.S. Army fighters who administer emergency trauma care on the battlefield. The contract funded HaptX and its partners Engineering & Computer Simulations and Mayo Clinic to integrate the true-contact haptics of HaptX Gloves with the Army's Tactical Casualty Combat Care (TC3) environment. For many years, TC3 used VR to teach soldiers to perform airway management, access vascular systems, apply tourniquets, etc. But without realistic touch feedback, trainees could only learn abstract procedural steps and could not practice key skills and physical interactions naturally. This contract funded research and development to improve the quality and retention of training to save more lives on the battlefield.

Figure 32: Soldier uses HaptX Gloves to interact with U.S. Army's Tactical Combat Casualty Care Simulation.

14.8.2. Customer Goals

Skills transfer, increased training effectiveness

14.8.3. Project Budget

>$100,000 USD

14.8.4. Haptic Technologies Used

The project incorporated multiple HaptX Gloves DK2 systems. These gloves maximize realism by combining high-fidelity tactile feedback, which displaces skin up to 2 mm, variable force feedback delivering up to 40lbs of resistive force per hand, and electromagnetic finger tracking.

14.8.5. Role of Haptics

The developers of this project applied haptic effects to the virtual interactions to enable trainees to physically feel and practice the training program's dexterous manual motions. These interactions included grasping and using supplies, feeling correct locations on the patient's body, turning the patient's body, and more. Haptics enabled the developers to achieve the project's core mission of advancing training through improved manual interfacing with the environment.

14.8.6. Outcome

The Army is midway through the project and will report its findings upon completion. Initial feedback has been positive.

15. Technology Options and Implementation Examples

15.1. Overview

This chapter provides additional use case examples with a specific focus on the technology and implementation choices used to create the simulations.

15.2. Military Training

Use Case Objective:

VR in defense and security is mainly used to improve the training of soldiers and officers and for the simulation of military missions and operations. The use of VR in military training has several benefits: it is a cost-effective alternative to military exercises. It can be carried out at any time, thanks to detailed reconstruction. The large number of possible combinations offered by VR offers unrivaled training breadth.

Haptics Objectives: The Haptics' objectives in military training are linked to *Realism*. The purpose of VR is to reduce the distance between real warfare and training.

Haptics Interactions are focused on realism. Some examples are:

- *Ambient Effects* to simulate the shocks of projectiles / environmental vibrations.
- *Clicks and dynamic controls* to simulate weapon gasket interaction.
- *Interactions (Hand-Object)* to manipulate weapons and interactable objects.
- *Active Objects (Fast)* like the recoil of a weapon.
- *Softness / Stiffness* for cardiac massage training.

Technology Options There is an ample spectrum of haptics technologies useful to enhance military training with different grades of complexity and costs. The most common and cost-effective haptic solutions are passive haptic props (e.g., weapons) that do not have any active haptic feedback. This approach has the value of passively providing a physical model with realistic inertial properties, whereas the dynamic behavior of the weapons like gasket force feedback and recoil are not included.

Better results are obtained with specifically built haptics-enabled weapon models like StrikerVR products, simulating both gasket force response and weapon recoil with a combination of vibrotactile and kinesthetic actuators. For the best simulation results, custom-built haptics-enabled dummy weapons should be used to replicate the dynamic behavior of the real weapons. This approach requires a more significant budget and custom development, but ensures the best simulative results.

To complement haptics-enabled weapons, haptics vests help immerse the user in the training activity with body located feedback simulating bullet shocks, vehicle vibrations, or environmental effects.

15.3. Procedural Training

Use case Objective

VR in training for skilled workers is a cost-effective way to train the workforce. VR training can be used for training new skills or improving existing knowledge. Haptics in VR training enables the worker to train procedures, interactions, and postures. VR training offers a better learning outcome than traditional video training because it is interactive. It is a more cost-effective solution compared to using real assets for training or dedicated workstations for educational purposes.

Haptic objectives for training for skilled workers should focus on two main items: skill transfer and realism.

Haptics Interactions:

- *Object manipulation: Manipulate objects with one or two hands.*
- *Shape:* Finding buttons, switches, dials, etc., on a dashboard or control panel.
- *Clicks and dynamic controls*: Pressing on a power tool trigger, operating buttons, sliders, etc. to control the environment.
- *Environmental Awareness:* Vibrations patterns indicating approaching danger, force-feedback from static objects in the scene to indicate physical boundaries.
- *Ambient Effects:* Vibrational patterns from a functioning piece of equipment.

Technology Options

The focus should be on skill transfer and can be done either by using gloves, force feedback devices, or by using an augmented passive tool/prop with tracking or combined solutions.

The advantage of gloves and force feedback devices is the ability to reconfigure the tool and interactions easily. However, the fidelity of the interactions will be lower than the tracked passive tool solution.

If realism is the primary objective, the augmented passive tool is likely a better solution path since it leverages all other aspects to that interaction (tool weight, material).

Vibrotactile haptics can be added to all these options to enhance the immersion of the overall experience. An example of that implementation will be the vibration of the working drill or a paint gun that vibrates when the paint flows from the nozzle or a glove that implements both force and vibrotactile feedback.

Tip From the Expert:

It depends on what you're training. If you're training something that needs a force-displacement relationship and it's not something that people have experience with, then you need that in haptics.

Dr. Allison Okamura, Stanford University

Regarding the cost and speed of the implementation, a more versatile tool such as the force feedback device might be a better choice in the long run, especially if the trainee needs to deal with many different tools in one task. If the task is very specific, an enhanced actual tool with vibrotactile is still an optimal solution and will allow users to focus on task performance.

The third option is a combined solution where a dummy tool is used, and the vibro-tactile or force-feedback glove is used to simulate less critical interactions in training. The gloves can also provide added feedback to the dummy tool.

15.4. Telerobotic

Use Case Objective

Tele-robotics is about controlling robotic arms and grippers from a distance. The aim of haptics in tele-robotics is to be able to perform tasks where human dexterity is required with a robotic set-up. Business objectives for such tasks could be to remove humans from dangerous or remote tasks. Another example of business objectives could also be to program complex robotic tasks easily.

Haptic objectives for tele-robotics aim to provide realistic, high fidelity to the operator to successfully perform (complex) dexterous tasks with the remote manipulators.

Haptics Interactions

- Object manipulation: Manipulate objects with one or two hands.
- Shape: Provide input to the operator on how the robot grips the objects.
- Environmental Awareness: Vibrations patterns indicating (potential) unwanted collisions with the robotic set-up.
- Ambient Effects: Vibrational patterns provide cues to the operator of objects that are sliding through the hand.

Technology Options

The choice of technology should mainly be influenced by the complexity of the task and the robotic end effector degrees of freedom. Humanoid grippers with many degrees of freedom are best matched with haptic devices that have similar degrees of freedom, e.g., a haptic glove. However, if only a two-fingered gripper is used, a less accurate haptic device might also be sufficient. The sensor fidelity at the end-effector is also a key consideration, and high-fidelity remote sensing will need to be matched with a similarly high-fidelity haptic actuation device. Kinesthetic feedback is best suited for telerobotic use cases, particularly when there is a spatial region within which the end effectors can move. Force feedback also enables the operator to understand the grip posture of the robot better. When tasks require dexterous manipulation of remotely grasped objects, skin deformation capabilities add to the user's capability of effectively manipulating the object within the hand. When two-handed interactions or environmental cues are also desired, a device, including vibrotactile feedback, is desired.

15.5. Virtual Prototyping

Use case Objective

Virtual prototyping involves using Virtual Reality technology to digitally design products and experiences.

Haptic objectives for virtual prototyping are linked to realism.

Haptics Interactions:

- *Object manipulation:* Manipulate objects with one or two hands.
- *Shape:* Finding buttons, switches, dials, etc., on a dashboard or control panel.
- *Clicks and dynamic controls*: Pressing on a power tool trigger, operating buttons, sliders, etc. to control the environment.

Technology Options

The most common haptics devices in virtual prototyping use cases are passive haptics elements (passive objects) to complement the virtual reality experience during the design phases. One example is for a virtual prototype for in-car HMI experiences where the HMI interface is a physically flat surface upon which the hand of the user in VR collides when interacting with the virtual screen. These implementations generate remarkable results with a limited investment. Their drawback is the static nature of such deployments.

One option to increase the interactivity of dummy objects is to leverage the concept of augmented haptics using a wearable haptic device. The digital haptics sensations created by the haptic device complement the touch sensations generated by the dummy object to create a blend of real and digital haptics experiences. This implementation increases the flexibility of the implementation, overcoming the limits of a static implementation with dummy objects. The challenge of this hybrid haptic approach is the need to guarantee excellent spatial registration between the passive surfaces and the active haptic elements to keep a consistent experience for the user.

15.6. Marketing

The difference between virtual and mixed reality is getting thinner and thinner. With the recent implementation from Oculus Quest of the passthrough mode, a few hundred-dollar device is capable of successfully creating mixed reality experiences like HoloLens and Magic Leap. In marketing, AR apps represent the most applicable use of such technologies. AR apps in marketing are usually used for engaging interaction between the consumer and the brand or the product. Showing the product is usually founded to be more interesting than only brand-related AR apps, and they present many advantages related to the accessibility to be anywhere at any time for customers. Recent examples show that virtual try-on or product-related AR apps increase customers' evaluation of products and brands, and particularly this is initiated by the playfulness and pleasurable moment spent with using the AR app. Overall, the experience drives the hybrid consumer experience rather than the utilitarian role of the AR app for product demonstration. Yet, the deep utilitarian value is still a requirement, but it must engage the consumer senses via vividness, interactivity, and 3D display. Adding haptic to this environment is under investigation, and recent work shows that it has positive experiential value for consumers needing touch stimulation to fulfill hedonic needs.

Haptic objectives for providing interactive and engaging hybrid experiences that enhance brand equity and increase online customer purchase via mobile application.

Haptics Interactions:

- *Object manipulation:* Manipulate objects with one or two hands.
- *Clicks and dynamic controls*: Pressing on a power tool trigger, operating buttons, sliders, etc., to control the environment.
- *Environmental Awareness:* Vibrations patterns indicating approaching the object or interacting with it, force-feedback from static objects in the scene to indicate physical boundaries.
- *Ambient Effects:* Vibrational patterns from a functioning piece of equipment.
- *Interactions (Hand - Object)*: Perception of an object being interacted with. For example, grab and move a lever, door, or slider.
- *Multisensory events:* Haptic Feedback in correlation with visual and audio cues to reinforce visual or audio objectives.
- *Objects Manipulation (Large)*: Perception of an object being manipulated. For example, the manipulation of a ball hold with the whole hand.

Technology Options

The most common haptics devices in XR marketing are the vibrotactile actuator in smartphones. The ubiquity of such actuators in the hands of every potential customer allows for a remarkable opportunity to leverage the multisensory experiences to increase conversion.

Most of the technologies presented in the book have a potential application in marketing; haptics gloves are useful to demonstrate the functioning of complex products, midair haptics to engage the user in MR scenarios.

The most challenging aspect of creating a successful XR haptics marketing application is to deliver a seamless experience to potential customers with low friction of usage and a delightful multisensory experience.

16. Conclusions

Since the democratization of virtual reality head-mounted displays from Oculus, one of the missing pieces for the virtual world is the sense of touch. This desire sparked a deep interest in academia and the startup ecosystem. Today, some of these startups are established players with incredible products and offering to bridge the gap between imagination and reality. Some of them contributed to writing this book.

This book gives a first-level introduction to the value of Haptics in extended reality from a practical point of view. It results from in-depth market feedback, insights, and the documentation of best practices and tips learned by veteran haptics professionals. Haptics can bring tangible value to real-world use cases today, even with its complexities and idiosyncrasies. The authors hope that this book helps bridge the information gap between developers and product managers and the haptics world.

The authors will strive to keep this material up to date in the following years to give a continuously updated guide for implementers.

17. Acknowledgements

This book is the product of the XR Haptics working group of the Haptics Industry Forum. The authors would like to thank all the contributors listed on the book cover and all the contributors with use cases and feedback.

A special thanks goes to Daniel Shor and Willian Firer for contributing to kickstart the XR Haptics group within the Haptics Industry Forum and gathering the first set of data.

Similarly, a special thanks goes to Dave Birnbaum from Expressive Foresight and Fabien Danieu from Interdigital for providing an in-depth review and feedback about the content.

18. Authors Biography

Eric Vezzoli is the CEO and haptics architect of Interhaptics, leading a team of engineers and product-marketing to launch the Interhaptics Haptics Composer: the first multiplatform Haptics Design tool on the market. He is the main author of the compression technologies behind the upcoming MPEG haptics standards.

Eric was the CTO of Hap2u in charge of the software team designing the rendering stack to drive surface haptics technologies at full capabilities. Hap2U raised $6M from Daimler.

He holds a PhD in Power Electronics and Haptics from INRIA and Lille University and a Master of Mathematical and Physical Modeling from Politecnico of Milan.

He published 20+ scientific papers on haptics and human-machine interactions and deposed 6 patents on haptics technologies and applications.

Chris Ullrich is the CTO of Immersion Corp., where he is responsible for ensuring that haptic products and technologies designed and developed at Immersion are highly valued by users, application developers, and OEM customers.

A prolific inventor with more than 90 issued US patents in human-computer interfaces, virtual reality, algorithms, novel actuation methods, user interfaces, and touch apps, Chris Ullrich is a key driver of innovation at Immersion.

Since graduating with a degree in Applied Mathematics from the University of British Columbia, he has worked in many areas of virtual reality, mobile devices, automotive, surgical instruments, and ad-tech.

In particular, Mr. Ullrich co-developed the CyberGlove / CyberGrasp whole-hand haptic exoskeleton for virtual prototyping, a CathSim trainer for minimally invasive cardiac catheterization procedures, and haptic technologies used in the PlayStation DualSense controllers. Mr. Ullrich is widely regarded as an expert in haptic system design and development with specific and extensive experience in XR applications.

Gijs den Butter is the co-founder and chief of product at SenseGlove - a Dutch company developing and commercializing SenseGlove. A wearable force and vibrotactile feedback glove that enables lifelike interactions in XR.

Gijs holds its master's in design engineering from the Delft University of technology. Gijs is specialized in human-computer interaction and entrepreneurship. Prior to founding SenseGlove, Gijs worked as a UX designer and consultant.

Rafal Pijewski is the co-founder and CTO of Actronika. He is passionate about new technologies, especially fascinated by haptics and all it can bring to our lives.

He has extensive experience in CAD, CAM, FEA design, manufacturing processes, and industrial standards. Together, these competencies enabled him to carry out projects of bringing new ideas in the real world, through the design and manufacturing, to ready products.

19. Glossary

Active Force Feedback: A type of kinesthetic feedback in which the user's motion is modified by generated force. This is often implemented with powered motors.

Active Object (fast): A haptic interaction in which an object changes its physical state quickly with respect to user interaction. Typically, > 5Hz.

Active Object (slow): A haptic interaction in which an object changes its physical state slowly with respect to user interaction. Typically, in the 0-5Hz range.

Allocentric: A sensation in which haptics play in response to user interaction and is responsive to user interactions.

Ambient Effects: A haptic interaction providing users with a global or environmental sensation, such as a passing train.

Clicks and Dynamic Controls: A haptic interaction that reproduces the sensation of a physical control such as a push button.

Collision Detection: A simulation process in which the proximity and intersection state of 3D objects is calculated at simulation rate.

Congruency: The quality or nature of haptic feedback when considering other modal feedback from audio and visual.

Consistency: A haptic goal that is concerned with the ability of the haptic system to create the same forces, vibrations, or other tactile stimuli from user to user and session to session.

Contextual Awareness: A haptic interaction providing users with haptics that give them cues about their context.

Descriptive Effect: A haptic effect represented using a device-independent description of its desired experience.

Dynamic Interactions: A quality of a haptic interaction in which the haptic feedback changes with the user's motion.

Dynamic Range: The effective range of stimulation that a haptic device can generate. For example, a vibration motor may generate mechanical vibration from 0.2 to 5 gravities of acceleration.

Egocentric: A sensation in which haptics play on a specific part of the user's body is generally independent of user motion or action once started.

Electrostimulation: A type of contact spatial feedback in which electrodes directly stimulate mechanoreceptors to generate nerve afferents.

Expressivity: A haptic goal concerned with the dynamic range of the haptic device, as experienced by the user.

Haptics: Technologies intended to provide touch feedback to users.

High-Definition Vibration Effect: A vibration effect that includes variance in both magnitude and frequency.

Immersion: A haptic goal concerned with a user's sense of presence or belief in the reality of an XR experience.

Internal Dynamics: Simulated objects with internal degrees of freedom or change in response to user interaction.

Kinesthetic Effect: An effect that describes a change in force as a function of a portion of the user's position or orientation.

Kinesthetic: Haptic feedback that provides a force on a user's body. This can be ground referenced or user referenced force.

Latency: The total time from a user's sensed motion until the user feels a haptic sensation.

Mechanoreceptors: Nerve endings responsible for sensing tactile stimulation.

Multisensory events: A quality of a haptic interaction in which additional modalities are simultaneously presented to the user.

Nociceptors: Nerve endings responsible for the sense of pain.

Non-Diegetic Events: Haptic sensation provided to the user outside the scope of the simulation.

Object manipulation (large): A haptic interaction related to the gross position/orientation of the user's hand(s).

Object manipulation (small): A haptic interaction related to individual fingers' dexterous position/orientation.

OpenXR: An industry-standard API that provides consistent interfaces for various XR hardware devices. The standard is developed and maintained by the Khronos Group.

Passive Haptics: The kinesthetic/vibrotactile sensation generated by a physical prop. A physical weapon or tool that a user holds during an XR simulation provides passive haptic feedback.

Proprioceptors: Nerve endings responsible for sensing the kinematics of the user (e.g., joint angles).

Pseudo Haptics: A haptic sensation perceived by a user due to a visual/proprioceptive discrepancy. A virtual hand constrained to the surface of a virtual object when a user can fully squeeze their hand provides a type of pseudo-haptic sensation of the object shape.

Realism. A haptic goal concerned with providing haptic feedback that is close as possible to the real world.

Resistive Force Feedback: A type of kinesthetic feedback in which the user's motion is impeded. Often implemented with a physical braking mechanism.

Robustness: A haptic goal concerned with the system's ability to be shipped, set up, and used with the target throughput without failure.

Safety: A haptic goal concerned with the level of risk associated with using the XR equipment for the target user population.

Self-Referenced Haptics: Haptics generated when a user collides with their own body.

Shape: A haptic sensation providing a virtual object's overall 3D shape profile.

Shock Interaction: A haptic interaction with a sharp or surprising sensation profile.

Skills Transfer: A haptic goal concerned with an XR simulation's ability to create net positive training metrics.

Skin Indentation: A type of contact spatial feedback where the user's skin is deformed locally. Often implemented using a pin array.

Softness/Stiffness: The quality of an object to resist squeezing or contact pressure.

Static Interactions: A quality of haptic interaction in which the haptic effect is always the same regardless of the user's state.

Surface Friction: A type of contact spatial feedback in which the friction of a surface is modulated as a user interacts with it. Often implemented using an electrostatic actuating means.

Synchronicity: The inter-modal latency between haptics, audio, and visual feedback.

Synthesized Dynamic Vibration Effect: A vibration effect that varies as a function of some portion of the user's position or orientation.

Tactile/Contact Spatial: Haptic feedback that modulates sensation through surface contact, such as a physical surface with controllable friction properties.

Tactile/Non-contact Spatial: Haptic feedback projected onto the user's body through the air.

Tactile/Vibration: Haptic feedback that includes a time-varying component in the 5-500Hz range.

Textures: A haptic effect that varies with the spatial motion of a user's hand.

Transient Effect: A haptic effect with extremely short duration and high amplitude. Typically used for clicks, shocks, and other confirmation-oriented haptic interactions.

Transparency: A haptic goal concerned with the haptic sensation created by a haptic device when it is not actively generating haptic feedback.

Ultrasound: A type of non-contact spatial feedback in which the tactile stimulation is generated with focused ultrasound arrays.

Unity: A game engine widely used for XR simulation development.

Unreal: A game engine widely used for XR simulation development.

Usability: A haptic goal concerned with the comfort and adaptability of the haptic hardware used in the XR simulation.

User Experience: A haptic goal concerned with the overall execution quality of the XR environment and the ease with which users can make sense of it.

XR/eXtended Reality: Extended Reality refers to digital experiences that incorporate at least one of virtual, augmented, or mixed reality.

20. Works Cited

[1] Haptics Industry Forum , [Online]. Available: https://hapticsif.org/

[2] Y. G. S. L. Y. Z. W. X. J. X. Dangxiao WANG, Haptic display for virtual reality: progress and challenges, Virtual Reality & Intelligent Hardware, Volume 1, Issue 2, 2019

[3] bHaptics, "https://www.bhaptics.com/," [Online]

[4] Actronika, "Skinetic," [Online]. Available: https://www.skinetic.actronika.com/

[5] Senseglove, [Online]. Available: https://github.com/Adjuvo/SenseGlove-Unity

[6] Haptx, "https://haptx.com/," [Online]

[7] dexmo, "https://www.dextarobotics.com/," [Online]

[8] Haption, "https://www.haption.com/en/," [Online]

[9] tanvas, "https://tanvas.co/," [Online]

[10] https://www.hap2u.net/. [Online]

[11] G. T. VR, "https://www.gotouchvr.com/," [Online]

[12] Weart, "https://www.weart.it/," [Online]

[13] Teslasuit, "https://teslasuit.io/," [Online]

[14] Ultraleap, [Online]. Available: https://developer.ultrahaptics.com/

[15] Emerge, "https://emerge.io/," [Online]

[16] Striker VR, [Online]. Available: https://www.strikervr.com/

[17] Protube VR, [Online]. Available: https://www.protubevr.com/en/

[18] S. K. Z. W. M. C. Razzaque, "Redirected Walking," 2001

[19] M. H. H. B. E. O. A. D. W. Mahdi Azmandian, "Haptic Retargeting: Dynamic Repurposing of Passive Haptics for Enhanced Virtual Reality Experiences," 2016

[20] Interhaptics, "VR Interactions Essential," [Online]. Available: https://assetstore.unity.com/packages/tools/utilities/vr-interactions-essentials-by-interhaptics-168830

[21] Auto Hand, [Online]. Available: https://assetstore.unity.com/packages/tools/modeling/auto-hand-vr-physics-interaction-165323

[22] Clap XR, [Online]. Available: https://clapxr.com/

[23] Interhaptics - Self-Referenced Haptics, [Online]. Available: https://youtu.be/2JEusvyZBHs

[24] Unity. [Online]. Available: https://docs.unity3d.com/ScriptReference/XR.HapticCapabilities.html

[25] Khronos Groutp, [Online]. Available: https://www.khronos.org/openxr/

[26] Interhaptics, [Online]. Available: www.interhaptics.com

[27] Audiokentic, [Online]. Available: https://www.audiokinetic.com/products/wwise/

[28] Ffmod, [Online]. Available: https://www.fmod.com/

[29] Actronika, [Online]. Available: https://www.actronika.com/solutions-for-realistic-haptics#unitouch-technology

[30] B. S. P. R. C. D. S. G. P. A. K. Sharon Oviatt, The Handbook of Multimodal-Multisensor Interfaces: Foundations, User Modeling, and Common Modality Combinations - Volume 1, Association for Computing Machinery and Morgan & Claypool, 2017

[31] [Online]. Available: https://www.uni-weimar.de/fileadmin/user/fak/medien/professuren/Virtual_Reality/documents/publications/2012-IEEE-3DUI-God-Hand.pdf

[32] F. Danieau, "HFX Studio: Haptic Editor for Full-body Immersive Experiences," [Online]. Available: http://fdanieau.free.fr/pubs/hapticEditor.pdf